Financing the Race t

Our financial imagination
is as important as our technological imagination
when it comes to extending our reach into the cosmos.
Armen V. Papazian
Starship Congress 2013
Dallas—Texas, USA

Armen V. Papazian

Financing the Race to Space

How to Value, Invest, and Explore the Universe

palgrave
macmillan

Armen V. Papazian
Space Value Foundation
London, UK

ISBN 978-3-031-73101-3   ISBN 978-3-031-73102-0   (eBook)
https://doi.org/10.1007/978-3-031-73102-0

Cover illustration: shulz

This Palgrave Macmillan imprint is published by the registered company Springer Nature Switzerland AG
The registered company address is: Gewerbestrasse 11, 6330 Cham, Switzerland

If disposing of this product, please recycle the paper.

Is the fulfilment of these ideas a visionary hope? Have they insufficient roots in the motives which govern the evolution of political society? Are the interests which they will thwart stronger and more obvious than those which they will serve?

… At the present moment people are unusually expectant of a more fundamental diagnosis; more particularly ready to receive it; eager to try it out, if it should be even plausible. But apart from this contemporary mood, the ideas of economists and political philosophers, both when they are right and when they are wrong, are more powerful than is commonly understood. Indeed the world is ruled by little else. Practical men, who believe themselves to be quite exempt from any intellectual influences, are usually the slaves of some defunct economist. Madmen in authority, who hear voices in the air, are distilling their frenzy from some academic scribbler of a few years back. I am sure that the power of vested interests is vastly exaggerated compared with the gradual encroachment of ideas. Not, indeed, immediately, but after a certain interval; for in the field of economic and political philosophy there are not many who are influenced by new theories after they are twenty-five or thirty years of age, so that the ideas which civil servants and politicians and even agitators apply to current events are not likely to be the newest. But, soon or late, it is ideas, not vested interests, which are dangerous for good or evil.

John Maynard Keynes
The General Theory of Employment, Interest and Money. Macmillan.
1936, 383–384

To dream and matter in space…

Foreword

It is my great pleasure to introduce Armen V. Papazian's ground-breaking work, 'Financing the Race to Space.' This book provides a visionary and well-crafted exploration of the monetary and financial frameworks necessary for humanity's pursuit of outer space development, exploration, and settlement.

Papazian's insights are both profound and timely, addressing the complex challenges of financing space endeavours while offering innovative solutions to unlock the investment programs required for future space habitats. His thorough analysis and forward-thinking propositions make this book an essential read for anyone interested in the future of space exploration.

The nuanced discussion highlights the importance of a cooperative approach to space exploration, advocating for a blend of competition and collaboration that can drive humanity forward. Papazian's arguments are compelling, grounded in deep research and presented with clarity and conviction.

As the 695th Lord Mayor of London I have been deeply involved in promoting space sustainability through the Space Protection Initiative. I find Papazian's emphasis on the need for innovative financial instruments, particularly relevant. These concepts align with our ongoing efforts to ensure that our ventures into space are both economically viable and environmentally responsible, particularly our efforts for space debris removal insurance bonds.

'Financing the Race to Space' is a masterful work that challenges and inspires. Armen V. Papazian's book is a must-read for policymakers, investors, and anyone with a keen interest in the future of space exploration. It is a testament to the power of visionary thinking and a significant contribution to the discourse on sustainable space development.

UK

June 2024

Alderman Professor Michael Mainelli

The Right Honourable The Lord Mayor

of London

Preface

Image Credit: NASA, Voyager 1, 1990, 3.7 bn miles (6 bn km) from the Sun.

"Look again at that dot. That's there. That's Earth. On it lived all those you will never meet and have never heard of, the species that ravaged its own home from under its own feet. In crafted confusion, they served imaginary debts and very real greed, failed to expand their reach, and suffocated themselves to oblivion–on a mote of dust suspended in a sunbeam."

Sarl Cagan

Alien Astronomer[*]

[*] This quote is by the author—a satirical warning inspired by Carl Sagan's famous paragraph in Pale Blue Dot (Sagan, 1994): "Look again at that dot. That's here. That's home. That's us. On it everyone you love, everyone you know, everyone you ever heard of, every human being who ever was, lived out their lives. The aggregate of our joy and suffering, thousands of confident religions, ideologies, and economic doctrines, every hunter and forager, every hero and coward, every creator and destroyer of civilization, every king and peasant, every young couple in love, every mother and father, hopeful child, inventor and explorer, every teacher of morals, every corrupt politician, every 'superstar,' every 'supreme leader,' every saint and sinner in the history of our species lived there–on a mote of dust suspended in a sunbeam."

I felt compelled to write this book. While some of the arguments I discuss have been shared in two previous publications, *The Space Value of Money* and *Hardwiring Sustainability into Financial Mathematics*, the main theme, propositions, and style of this work are different and unique. I have also updated some of the concepts and included new material to make the arguments clearer and more relatable. Unlike the first two, both written for professionals and academics in finance, this book is designed to share the key insights of my research with a much wider audience.

I must clearly state from the outset that this book is entirely based on my own work and writing. I have not used any AI (Artificial Intelligence) tool. This is important and relevant for two main reasons. The first one is to address broader intellectual property concerns and to ensure the reader is aware that the words in this book have not been chosen through a series of probabilistic decisions. I have selected them based entirely on their ability to express the insights that preceded them. When necessary, words have been replaced and refined, sentences have been improved, to embody the ideas in their most authentic form.

This brings me to the second more philosophical reason. In my view, the most critical challenge of Generative Pre-Trained Transformers (GPT) and the Large Language Models (LLMs) they rely on is the asymptotic and probabilistic nature of their intelligence. As such, however vast their data sets, however fast their processors, however extensive their training, however precise their memory, and however functionally and mechanically useful the machines may be, we are the ones who must invent the new equations upon which our improved future will be built. We must feed the machines, and indeed we do, directly and/or indirectly, consciously, or not.

I am not in a position to venture a guess regarding the timelines and capabilities of Artificial General Intelligence (AGI), but I can comfortably state that, for the foreseeable future, the purposeful, intentional, and creative application of human imagination will continue to be the driving force of human progress. This may change with the arrival of artificial intentional imagination (AIIM).

While AI is a defining element of our current and future progress, and relevant to the technological advances we will achieve on our way to the stars, it is not a core theme of this discussion. In other words, the above paragraphs are the extent of everything AI related in these pages.

This book lays down the monetary and financial foundations for a species actively pursuing outer space development, exploration, and settlement and offers the keys to unlocking the massive investment programs needed to invent, manufacture, deploy, and maintain the new habitats of the future.

The reference to the 'race to space' in the main title aims to contextualise the subject for a wider audience. Personally, I believe humanity's outer space expansion should be a coopetition, i.e., a form of cooperative competition, not a race. Interestingly, the title can also be read to mean 'financing the *human race* to space,' which is exactly what this book is about. Furthermore, although the title refers to space, this book is about achieving meaningful milestones in *outer space*. This is an important nuance given that I define space as our physical context of matter, irrespective of constitution, composition, density, dynamics, and temperature, stretching from subatomic to interstellar space and every layer in between and beyond, where outer space, however vast, is but a segment.

In a molecular field of matter (and energy), where our imagination is a primary force of motion and transformation, everything we do and create on planet Earth is a direct function and result of our interpretations of ourselves and the universe. Our bodies, our planet, our solar system, the Milky Way, and the entire observable universe are part of the space within and through which we act upon interpretation. Thus, what we do, invent, and achieve in this vast molecular context comes down to our own creative imagination, and our willingness to work and sacrifice for its realisation. This is true for all of us, individually and as a global collective. While I do not delve into philosophical discussions on the nature of reality, this is an important starting point as it determines the realm of what we consider possible.

Wherever one may be on planet Earth, or on the International Space Station (ISS), or Tiangong Space Station (TSS), at this very moment, and always, one is in between stars—above one's head and underneath one's feet. This is the true and authentic *where* of the human experience—a tangible given, more real than the entire taxonomy of our projected beliefs, whether economic, financial, monetary, or other. Indeed, however limiting our interpretations of our reality may be, whether on the level of the individual, a collective, or humanity as a whole, our context suggests otherwise.

Sandwiched between stars, to explore the universe our technological imagination must be matched with a commensurately bold and empowering financial imagination. We transcend our limiting interpretations in technology every day, and it is high time we find the new concepts through which we can address the debilitating assumptions of our financial and monetary economics.

At the edge of an ecological catastrophe, on Earth, in space, we must find a way to transform and remove the intellectual and structural impediments to our evolution and the survival of our children. Ultimately, our struggle is not with carbon in the air, with plastic in the oceans, with sewage in the rivers, or with the debris in orbit. It is also not with gravity. Our truest challenge is human mediocrity, in all its shapes and forms, in ourselves and around us, reducing our planet into a consumable of our own confusion, and the universe into a failed version of itself.

Indeed, it is entirely possible that there are other more enlightened civilisations in the universe who have no need for money and do not require monetary incentives to guide their own productive power. It is also entirely possible that they do not need to be persuaded to value the ecosystem they inhabit, and thus, they do not need to be stopped from destroying their own home. As such, from their perspective, the entire content of this book can be considered a worthless evolutionary compromise.

As individuals, nations, and humanity, we are in the grip of our own interpretations and misinterpretations, and we must rethink and reimagine ourselves to improve the world and extend our reach. This book aims to offer an alternative path that can empower us with the tools and concepts that can achieve both, our sustainability on Earth and expansion in outer space.

UK   Armen V. Papazian
July 2024

Acknowledgements

I owe thanks and gratitude to a number of institutions and individuals who have supported this project. Given that this work is partly based on two previous publications, *The Space Value of Money* and *Hardwiring Sustainability into Financial Mathematics*, the acknowledgements and support I have received for both books are relevant to this publication as well.

I am grateful to:

City of London Corporation for supporting my work and taking part in the launch of *Handwriting Sustainability into Financial Mathematics*.

King's College Cambridge University for continuously inspiring my work and partnering in the launch of *The Space Value of Money* and *Handwriting Sustainability into Financial Mathematics*.

Judge Business School Cambridge University for giving me the platform to express the early insights discussed in this book and partnering in the launch of *The Space Value of Money* and *Handwriting Sustainability into Financial Mathematics*.

The Chartered Institute for Securities and Investment for supporting my work and sponsoring the launch of *Handwriting Sustainability into Financial Mathematics*.

Palgrave Macmillan for continuously supporting my work, ensuring its worldwide distribution, and partnering in the launch of *Handwriting Sustainability into Financial Mathematics*.

Federated Hermes Limited for sponsoring my research during the writing of *The Space Value of Money* and the launch event of *Handwriting Sustainability into Financial Mathematics*.

Institute of Chartered Accountants in England and Wales for hosting and supporting the launch of *Handwriting Sustainability into Financial Mathematics*.

Tech Nation for providing me with the sponsorship that allowed me to continue working on this book as well as *The Space Value of Money* and *Handwriting Sustainability into Financial Mathematics*.

National Space Society for endorsing an earlier paper on the topic and supporting my work.

Space Renaissance International for endorsing an earlier paper on the topic and inviting me to lecture on some of the key ideas discussed in this book.

18th SDG for Space Coalition for endorsing an earlier paper on the topic.

Space Value Foundation for continuously supporting my work and ideas, and for advocating for their adoption.

I owe special thanks to the 695th Lord Mayor of London, Alderman Professor Michael Mainelli, for accepting the invitation to write a foreword. I am inspired by his words and grateful for his support.

I owe special thanks to Dr. Gillian Tett, OBE, Karlton D. Johnson, Tracy Vegro, OBE, and Adrian Webb for reading an early draft of the book and providing reviews.

I owe special thanks to Prof. Gishan Dissanaike, Lt Col. Peter Garretson, Dr. Keith Carne, Dr. Saker Nusseibeh, CBE, Dr. Pascal Blanqué, Daud Vicary, Domenico Del Re, George Littlejohn, Dr. Matteo Cominetta, Eoin Murray, Giotto Castelli, Prof. Christine Hauskeller, Prof. Aram A. Amassian, and Dr. Jonathan Bonello for their support, for reading and providing reviews.

I owe special thanks to Tula Weis for her unwavering support, she has played a vital role in bringing this book to light, and to Susan Westendorf, Geetha Chockalingam, and Melvin Lourdes Thomas for a swift and highly professional production of the book.

I am grateful to the following individuals for their direct and/or indirect contributions, recently or in the past:

Prof. Dame Sandra Dawson, Dr. Mark Carney, Dr. Rob Wallach, Prof. Geoff Meeks, Prof. Arnoud De Meyer, Prof. Ha-Joon Chang, Dr. Robin Chatterjee, Dr. Jose Gabriel Palma, Prof. Tony Lawson, Prof. Geoffrey Hodgson, Prof. Pierre-Charles Pradier, Prof. Peter Nolan, Prof. Shailaja

Fennell, Prof. Richard Barker, Dr. Rachel Armstrong, Dr. Richard Obousy, Kelvin F. Long, Dr. Ian J. O'Neill, Amalie Sinclair, Alan Sinclair, Adriano Autino, John Lee, Prof. Pier Marzocca, Prof. Dirk Schulze-Makuch, Prof. Joseph Miller, Charles Radley, Giorgio Gaviraghi, Eric Klein, David Brin, Prof. Edward Guinan, Dr. Cathy W. Swan, Prof. Peter A. Swan, Dr. Jose Cordeiro, Dr. Robert L. Frantz, Prof. Weilian Su, Dr. Mae Jemison, Dr. David Livingston, Michael Laine, Dr. Eric Davis, Marc G. Millis, Paul Gilster, John Davies, Dr. Andreas Hein, Robert Swinney, Patrick Mahon, Robert Kennedy III, Angelo Genovese, Prof. Gregory Matloff, Tony Manwaring, Mary Priddey, Charles Goldsmith, Erin Hallett, Sandie Campin, Ruth Newman, Jane Kemp, Jane Playdon, Myra MacMillan, Maris Kraulins, and Dr. Ivan Collister.

I am also grateful to the many colleagues, students, friends, and family who have contributed to the wealth and depth of my learning and experiences over the years. I have learned so much from so many.

While all must be thanked, mistakes remain my own.

Reviews

The Space Value of Money
"Every once in a while, a book comes along that makes a fundamental contribution that is both profound and practical. A book that every member of the National Space Council, including the NASA Administrator and the Space Force Chief of Space Operations should read. *The Space Value of Money* will be of interest to ESG and impact investors, government regulators, financial theorists, and outer space enthusiasts."

—Lt Col Peter Garretson, *Senior Fellow in Defense Studies, American Foreign Policy Council*

"No doubt, the pressing environmental challenges we face make the concept of the space impact of investments even more compelling."

—Dr. Pascal Blanqué, *Chairman of Amundi Institute, Former Group CIO of Amundi Asset Management*

"*The Space Value of Money* brings much needed conceptual rigour, whilst further advocating the case for a new paradigm shift in financial valuation. This work gives us the lasting frameworks that aggregate impact across all spatial dimensions. Dr. Papazian culminates over ten years of research in this rich book, providing the springboard for further innovation and system implementation in this area."

—Domenico Del Re, *Director, Sustainability and Climate Change, PwC*

"Enthralling and captivating. Papazian offers a clear, thorough, and comprehensive discussion. The Space Value of Money gives us an opportunity to reframe our thinking and to explore what is possible. A great read!"

—Daud Vicary, *Founding Trustee of the Responsible Finance and Investment Foundation*

"Armen has developed a novel way to create financial models that are better suited to dealing with the many parameters required if we are to properly consider environmental factors and sustainability in economics and finance. I have found this engaging and look forward to seeing its future use."

—Dr. Keith Carne, *First Bursar, King's College, Cambridge University*

Hardwiring Sustainability into Financial Mathematics
"Dr. Papazian's *Space Value of Money* concept addresses sustainability in microeconomics and macroeconomics—it critically updates and accounts for the additional dimension. This should be compulsory reading for all students of finance and investing."

—Eoin Murray, *Head of Investment, Federated Hermes Limited*

"At a time when the climate crisis drives home the point for urgent action, and ESG measurements have come under intense scrutiny, one can hardly overstate the importance of this book. Rigorous and comprehensive, it offers an investment impact measurement methodology and answers the key financial question of our times: how can we fund the transition to a sustainable world? A finance handbook for the future."

—Dr. Matteo Cominetta, *Head of Macroeconomic Research, Barings LLC*

"In the bewildering and ever-swelling sea of acronyms that now covers the world of sustainable finance, Dr. Armen Papazian's laser-like focus on the financial mathematics of investment value and return is a most welcome addition to the profession's navigation skills."

—George Littlejohn, *Senior Adviser, Chartered Institute for Securities & Investment*

"This is a brave book—it highlights inconvenient truths about the financial mathematics of investment that guides the global flow of capital and offers thought through solutions. It is a must read for anyone concerned with planetary sustainability."

—Adrian Webb, *Founder & Director, Space Value Foundation*

Financing the Race to Space

"*Financing the Race to Space* is a masterful work that challenges and inspires. Armen V. Papazian's book is a must-read for policymakers, investors, and anyone with a keen interest in the future of space exploration. It is a testament to the power of visionary thinking and a significant contribution to the discourse on sustainable space development."

—Alderman Professor Michael Mainelli, *The Right Honourable The Lord Mayor of London, City of London Corporation*

"*Financing the Race to Space* is an inspiring and ground-breaking blueprint for the future of space exploration. Blending visionary thinking with practical financial innovation, it offers transformative insights into funding sustainable space ventures and provides the tools and guidance needed to propel humanity's most ambitious space endeavours. Papazian redefines the financial paradigm necessary for our leap into the cosmos. Essential and enlightening, a must-read for C-suite executives, investors, and policymakers."

—Karlton D. Johnson, *CEO and Chairman of the Board of Governors, National Space Society*

"Investors, wisely advised by our members who understand the precepts that Dr. Papazian covers so well in his work, stand to reap the responsible rewards."

—Tracy Vegro OBE, *CEO, The Chartered Institute for Securities & Investment*

"An extraordinary book that serves us and future generations with realism and optimism."

—Adrian Webb, *Founder and Director, Space Value Foundation*

"A thought provoking and fresh way to look at finance and economics with a set of ideas that deserve wider debate."

—Dr. Gillian Tett OBE, *Provost, King's College Cambridge University, Editorial Board Member, Financial Times*

Contents

About the Author

Armen V. Papazian is a financial economist and author of two books on sustainable finance, *The Space Value of Money* and *Hardwiring Sustainability into Financial Mathematics*. He is a founder and director of the Space Value Foundation. A Doctor of Financial Economics from Cambridge University, Judge Business School, King's College Cambridge, he is an active contributor to the public debate on sustainability in finance. A former investment banker, stock exchange executive, academic, and consultant, Armen combines extensive industry experience with financial institutions and markets, with in-depth research into the theoretical and practical aspects of sustainable finance. He is the first winner of the Alpha Centauri Prize for his work on money mechanics for space presented at the Starship Congress in 2013, in Dallas, TX, USA.

Abbreviations

AGI	Artificial General Intelligence
AI	Artificial Intelligence
AIIM	Artificial Intentional Imagination
APF	Asset Purchase Facility
APT	Arbitrage Pricing Theory
BEA	Bureau of Economic Analysis
BEAPFF	Bank of England Asset Purchase Facility Fund Limited
BOE	Bank of England
CAA	Climate Ambition Alliance
CAPEX	Capital Expenditure
CAPM	Capital Asset Pricing Model
CCL	Commerce Control List
CDO	Collateralised Debt Obligations
CE	Credit Easing
CGFI UK	Centre for Greening Finance and Investment
CGFI-SFI UK	Centre for Greening Finance and Investment, Spatial Finance Initiative
COPUOS	United Nations Committee on the Peaceful Uses of Outer Space
CSRD	Corporate Sustainability Reporting Directive
DBEIS	Department for Business, Energy & Industrial Strategy
DCF	Discounted Cash Flow
DDM	Dividends Discount Model
DDTC	Directorate of Defense Trade Controls
DOJ	Department of Justice
DSIT	Department of Science, Innovation & Technology

EA	Environmental Agency
EAR	Export Administration Regulations
ECB	European Central Bank
EEA	European Environment Agency
EFSF	European Financial Stability Facility
EO	Earth Observation
ESA	European Space Agency
ESG	Environmental, Social, and Governance
EU	European Commission
FCF	Free Cash Flows
FCFE	Free Cash Flows for Equity
FCFF	Free Cash Flows for Firm
FDR	Franklin D. Roosevelt
FED	Federal Reserve
GDP	Gross Domestic Output
GEO	Geostationary Orbit
GHG	Greenhouse Gas
GIS	Geographic Information Systems
GLBA	Gramm-Leach-Bliley Act
GO	Gross Output
GPGP	Great Pacific Garbage Patch
GPS	Global Positioning System
GPT	Generative Pre-Trained Transformers
GSA	Glass-Steagall Act
GSV	Gross Space Value
HRC	Habitat Replacement Costs
IEA	International Energy Agency
IMF	International Monetary Fund
IPBES	Intergovernmental Science-Policy Platform on Biodiversity and Ecosystem Services
IPCC	Intergovernmental Panel on Climate Change
IRR	Internal Rate of Return
ISSB	International Sustainability Standards Board
ITAR	International Traffic in Arms Regulations
LCA	Life Cycle Assessment
LEO	Low Earth Orbit
LLM	Large Language Models
MBS	MORTGAGE-BACKED Securities
MEO	Medium Earth Orbit
MOD	Ministry of Defence
MTSCPCP	Minimum Temporary Survival Condition for a Planet-Consuming Parasite
NASA	National Aeronautics and Space Administration
NHS	National Health Service

NOSD	New Outer Space Deal
NOSS	National Outer Space Strategy
NPV	Net Present Value
NSC	Nihilistic Survival Condition
NSS	National Space Strategy
NSV	Net Space Value
OECD	Organisation for Economic Co-operation and Development
OSC	Office of Space Commerce
OST	Outer Space Treaty
PA	Paris Agreement
PB	Planetary Boundaries
PCN	Public Capitalisation Notes
PM	Particulate Matter
PMCCF	Primary Market Corporate Credit Facility
PNT	Positioning, Navigation and Timing
PPP	Public Private Partnership
PRI	Principles of Responsible Investment
QE	Quantitative Easing
SF	Sustainable Finance
SI	International System of Units
SMCCF	Secondary Market Corporate Credit Facility
SVoM	Space Value of Money
TSS	Tiangong Space Station
UK	United Kingdom
UKSA	United Kingdom Space Agency
UNFCCC	United Nations Framework Convention on Climate Change
UNOOSA	United Nations Office for Outer Space Affairs
US	United States
USML	United States Munitions List
VE	Value Easing
WACC	Weighted Average Cost of Capital
WC	Working Capital

List of Figures

List of Charts

List of Tables

1

Introduction

I see Earth! It is so beautiful.
Yuri Gagarin, Vostok 1 Cosmonaut, 1961

Why don't you just fix your little problem and light this candle.
Alan Shepard, Mercury Freedom 7 and Apollo 14 Astronaut, 1961

"We humans are the greatest of Earth's parasites." This quote, attributed to Dr. Martin H. Fischer (1879–1962), a former professor of physiology at the University of Cincinnati, is obviously meant as a metaphor of our reality on the planet.[1] This book is about outer space exploration, development, and settlement, and as such, it is built on a vision of humanity that aims to take us beyond a parasitic interpretation of human civilisation.

I discuss the monetary and financial foundations necessary for unlocking the massive investment programs needed to invent, manufacture, deploy, and maintain the new habitats of the future. The main focus is on the essential transformations required to unleash our potential in outer space. If the main insights and propositions of this work are accepted and implemented, I believe they will create numerous opportunities that could lead to significant private benefits to outer space investors and entrepreneurs. Moreover, they can help us create the tools through which we can access the diverse

---

[1] Parasites are organisms that feed on and live in or on another organism. They grow and multiply, they harm their host, but need it to survive. Some parasites can also have beneficial impacts on their host.

© The Author(s), under exclusive license to Springer Nature
Switzerland AG 2024
A. V. Papazian, *Financing the Race to Space*,
https://doi.org/10.1007/978-3-031-73102-0_1

resource wealth that exists beyond our atmosphere. Furthermore, if nation-states recognise the far-reaching implications of the offered solutions, they can help themselves and achieve historic breakthroughs on their journey to the moon and Mars.

Ultimately, if the key insights and suggestions are adopted, we will be able to lawfully create and allocate the resources we need to achieve our most daring visions in outer space, while securing the sustainable continuity and expansion of human productivity on this planet and beyond.

Thanks to the awe-inspiring achievements of public agencies and private companies, like NASA and SpaceX, outer space has been in the news more often than ever before. The number of books, articles, social and traditional media publications, and professional industry reports keeps growing exponentially. 'Space' has become an integral part of our daily economic narrative. Indeed, we now have a fast-growing 'space economy.'

You may ask, what is the value proposition of this book if there is already a fast-growing 'space economy'? Well, it is my view that the current 'space economy' *cannot* and *will not* take us very far in outer space, not any time soon at least. I say this while fully acknowledging the splendid work done by public and private entities in the sector.

Let me clarify. The above does not refer to the technical and technological aspect of our journey in outer space. I am not an engineer or a physicist, and thus, cannot offer anything of value in that respect. The above statement concerns our financial and monetary reality and the funding structure of the sector.

As the key driving force behind all our investments, capital, public or private, underpins all our productive activities. Where, how, and why we invest are key questions that guide and direct our productivity. In other words, the guidance system of our investments, which defines our answers to where, how, and why we invest, has evolutionary implications. Naturally, not all investments achieve their intended results, and innovation is primarily possible thanks to our courage to dare, our ability to persist, and determination to keep going after failures. We cannot break new ground without continuous investments into that which we do not yet know or understand.

Indeed, beyond those questions, at a much deeper level, are the ones that concern money itself. Where and how we create money are equally relevant to our challenge. As such, our financial and monetary framework and reality are directly relevant to our future in outer space. They are key for unlocking the investment programs that can help us invent, manufacture, deploy, and maintain the new habitats beyond our atmosphere. This is what the above claim refers to, and what this book is about.

My purpose is to unravel the inner workings of our current reality and make purely 'mechanical' observations. I abstain from moral judgements, and I do not argue for or against any cause, theory, or system. I reveal and address the shortcomings of our current financial value framework, financial mathematics, and monetary architecture, discussing their implications for our ability to extend our reach in outer space.

Much of the recent enthusiasm and activity in the new 'space economy' are about using outer space infrastructure to deliver goods and services on Earth. There are, of course, many and great benefits to this momentum and the products and services it facilitates. From Earth observation and satellite communications to positioning, navigation and timing (PNT), the benefits are numerous. Nevertheless, the sector's absolute and relative size, thematic and industry focus, and funding structure reveal a 'space economy' that is still Earthbound and relatively small.

I discuss the key features and Earthbound nature of the sector in Chapter 2. A few words on the private and public parts of the outer space sector are necessary here. The private outer space sector is by default dependent on Earthly money supply. Unlike the public sector, which does not expect or seek any monetary return for its investments, the private outer space economy must generate cashflows and distribute and/or promise profits—that is its reason for being and how it survives as a going concern.

Given that the entire supply of money, in all its forms, is on Earth, and given that all actual and potential paying clients are on Earth, the non-negotiable imperative to generate cashflows constrains the private outer space sector and its activities. The necessity to seek and serve customers on Earth is imposed by the financial frameworks and principles that govern our markets and investments, which have primacy over any outer space exploration and settlement objective and/or achievement. Indeed, these constraints would not exist if private outer space sector firms could find customers and/or investors who would pay/invest in them to go to Mars just for the benefit of getting to the Red Planet, i.e., without expecting anything in return, now or in the future, in the form of goods, services, profits, or any other kind of benefits.

Public agencies are free from such constraints, they do not need to make a profit and do not expect monetary returns for their investments. However, they face their own set of limitations linked to the nature of their funding. Governmental budgets, and the debts they are often linked to, impose various kinds of restrictions on the public outer space sector.

In other words, the public and private parts of the outer space sector face structural constraints. The private outer space sector must pursue Earthly money supply to survive; the public outer space sector must live within the limits imposed by public budgets and growing debts. These limitations are themselves a function of our financial value framework, financial mathematics, and monetary architecture. Before elaborating this argument further, another important clarification is due.

You may have noticed that I referred to the *outer space* sector. This is because I do not use 'space' and 'outer space' interchangeably. I define space as our physical context of matter, irrespective of constitution, composition, density, dynamics, and temperature, stretching from subatomic to interstellar space and every layer in between and beyond, where outer space, however vast, is but a segment (Papazian 2022, 2023). As such, in my view, the new 'space economy' should be called the new *outer space* economy. I explore and discuss the need for a more precise terminology in Chapter 3.

This is an important nuance that offers a unique conceptualisation of our physical context, i.e., of space, with benefits that go far beyond terminological clarity. Just like many other conceptual lines that help us make sense of our physical context of matter, like the prime meridian, the equator, and the Kármán line, the layered conceptualisation of space is a foundational element of the tools and solutions I offer in later chapters.

Indeed, we have a serious civilisational challenge when it comes to our attitude towards space, as defined above. Chapter 4 reveals that humanity's treatment of space, including all its many layers, leaves a lot to be desired and needs serious examination and rethinking. After all, how can we terraform a non-habitable planet we may one day land on when we cannot stop ourselves from deforming the habitable planet we have got. Surely, we must recognise that addressing the immediate challenges we face here on Earth is of primary importance for our continuous survival and evolution.

In other words, our attitudes towards space must be addressed on a broader and more holistic level. How we treat space in general, Earth and outer space included, is where the discussion must begin. Of course, human civilisation may also remain on its current course and venture out into the cosmos as a planet-consuming space-littering parasite. While this may be possible, given our current mode of operation and impact on space, it will be a short-lived venture, and highly unlikely to lead to anything that could be qualified as a success. Thus, I believe our sustainability must be operationalised in space, understood in its broadest sense, considering the entirety of our physical context.

Sustainability does not mean 'degrowth' and/or ESG.[2] Yes, our planet and species are in turmoil, and we need fundamental solutions to the many challenges we face; but degrowth and ESG do not provide them. From climate change to waste and pollution, to ecological and socioeconomic crises, to wars and conflicts, we are suffering the consequences of many misinterpretations of our own making.Nestled in a vast cosmic landscape, on Earth, in space, our productivity is ruthlessly and relentlessly consuming our only home from under our own feet, while we struggle to invest and expand our reach in outer space.

What we are facing is a serious configuration challenge in the way we structure and incentivise human productive power. Instead of an inward drive that consumes the planet, we need to reconfigure ourselves and our productive systems for an outward adventure of discovery. This outward reconfiguration of human productive potential is essential for both, addressing our environmental and sustainability challenges on Earth and powering our expansion in outer space.

I argue that *our inability to invest and build a sustainable and fair reality on Earth* and *our inability to invest and expand our reach in outer space* are intricately related. They are both the consequence of the same theoretical, mathematical, and structural omission. They are both directly caused by our current financial value framework, financial mathematics, and monetary architecture. Thus, the reconfiguration I propose concerns the fundamental assumptions and omissions of our financial and monetary economics. The guidance system of investments and the very logic of money creation are the foundations that shape the above limitations.

The fact that all money supply is on Earth, and thus money chasing private outer space entities are bound to Earth, is only one part of the challenge. The other parts, which also affect the public outer space sector, have to do with the very framework, principles, and equations of finance, and the very nature and core assumptions of our monetary architecture. I discuss them in parts 2 and 3, Chapters 5 to 10.

As shocking and surprising as this may be for some, to date, space and outer space are missing from our financial value framework, financial mathematics, and monetary architecture. Our analytical framework in finance is entirely built around risk and time, without space, and serves only one stakeholder, the *mortal risk-averse return-maximining investor*.

---

[2] Environmental, Social, and Governance factors. ESG is a popular acronym and framework though which many believe they can address our sustainability related challenges. I disagree. I do not discuss the currently popular ESG-based sustainability standards and frameworks in this book. However, you can find a detailed critique in Papazian (2022) and (2023).

The abstraction of space, as analytical dimension, and our physical context, implies that planet and humanity are left out of our models—treated as a voluntary optional addendum, a side discussion. The abstraction of space from our framework has led to equations that do not consider the space impact of investments as relevant. Moreover, along with the dimension of space, they omit our responsibility for space impact. In other words, our current predicament and civilisational disrespect for space can be traced back to our financial value framework, mathematics, and monetary architecture.

Furthermore, the two principles of value upon which the current analytical framework of finance is built, Risk and Return and Time Value of Money, mirror and explain the focus on risk and time. These principles establish a unique logic in our equations and valuations. High risks and distant returns are negatively priced; the mortal risk-averse return-maximising investor prefers safe dollars today over risky dollars tomorrow. In other words, our current framework is inherently incapable of addressing the distant returns and very high risks involved in outer space exploration, development, and settlement projects. Indeed, the current principles of finance discriminate against all our evolutionary investments.

This book is not just a diagnostic review of our current reality. While I do expose our excessive reliance on and attachment to risk and time, the conceptual and theoretical ball and chain that we must transcend in order to unleash our potential in outer space, I also offer the blueprint of a new financial and monetary framework that entrenches respect for space, integrates the value of space, and empowers us on our journey to a multi-habitat future in outer space.

The framework and tools I propose can lead us to a space-adjusted conceptual framework that provides the necessary theoretical, mathematical, structural, and institutional solutions needed to address both of the above-identified challenges: our inability to invest and build sustainably on Earth and our inability to invest and expand our reach in outer space. The proposed solutions allow us to create and invest the necessary resources to sustainably finance our footprint in space, a concept that I describe as the *monetisation of space*.

In Part 4, Chapters 11 to 14, I provide the theoretical and mathematical solutions necessary to unlock our spatial potential and trigger our next big leap in outer space. You do not need to be a mathematician to understand and apply the equations, and they are explained in plain English in the text. You can also understand the ideas and concepts without the equations if you choose to.

In part 5, Chapters 15 to 17, I address the side benefits of a space-adjusted framework in relation to two actual case examples: a US wealth floor replacing a debt ceiling, and a UK New Outer Space Deal replacing years of austerity and underinvestment. These applications demonstrate how the suggested framework and solutions can be translated into policy and applied to address fundamental economic and institutional challenges. In truth, it is my view that the space value framework is key to transcending the institutional and policy matrix that has led us to our current predicament. In theory and in practice, the discussion is relevant to all countries. I have chosen the US and the UK as examples.

The reconfiguration proposed in this book builds upon what we already have. It does not require us to reinvent the wheel, and it does not imply a radical destruction of existing systems. It also does not encourage protests for or against some or other vested interest group. Echoing John M. Keynes, "I am sure that the power of vested interests is vastly exaggerated compared to the gradual encroachment of ideas" (Keynes 1936, 384).

What I offer is an improvement that can be achieved with minimal systemic shocks, leading to a win–win solution that frees us from the limiting interpretations that thwart our evolution in outer space and threaten the survival of our children on Earth.

The rest of the book is divided into 5 parts and 16 Chapters.

PART 1: Economy, Space, Outer Space, and Human Impact—introduces the 'space economy' and its main features, defines space and outer space, and explores different aspects of human impact on space.

Chapter 2: The New Outer Space Economy—introduces the new outer space economy revealing key features of the private and public components of the sector.
Chapter 3: Space & Outer Space—introduces the distinction between space and outer space and offers a foundational conceptualisation of our physical context, of space.
Chapter 4: Human Impact on Space—provides a snapshot of our impact on space, revealing widespread and brazen disrespect of space and its many layers.

PART 2: The Chains of Risk and Time—discusses our current financial value framework, financial mathematics, and monetary architecture, revealing key shortcomings, and the limitations they impose on our ability to invest and build sustainably on Earth and our ability to invest and expand in outer space.

Chapter 5: Spaceless Financial Value Framework—introduces and discusses our current spaceless financial value framework, built entirely around risk and time, based on two principles of value, risk and return and time value of money.

Chapter 6: Spaceless Financial Mathematics—introduces and discusses our current financial mathematics, specifically our equations of value and return in finance, and reveals key structural shortcomings: the absence of space and the omission of space impact from our models.

Chapter 7: Spaceless Debt-based Monetary Architecture—introduces the debt-based nature of our monetary architecture and identifies three structural bottlenecks of debt-based money. It also argues that cryptocurrencies are equally spaceless and not a viable alternative.

PART 3: Muzzle, Leash, and Whip in Space—discusses the three bottlenecks of our debt-based monetary architecture, revealing their stifling and inhibiting impact on our ability to invest and create a sustainable reality on Earth and our ability to invest and expand in outer space.

Chapter 8: Calendar Time: A Muzzle in Space—discusses the first structural constraint created by our monetary architecture and reveals the limitations of the social construct in relation to our outer space ambitions.

Chapter 9: Monetary Gravity: A Leash in Space—discusses the second structural constraint created by our monetary architecture and reveals the limitations imposed on our ability to expand in outer space.

Chapter 10: Monetary Hunger: A Whip in Space—discusses the third structural constraint imposed by our monetary architecture and reveals the downside and implications of using debt as a foundational logic of money creation.

PART 4: Breaking Risk and Time—offers the necessary solutions, practical steps, and mechanisms through which we can unlock the massive investments necessary for our expansion in outer space. It upgrades and expands our financial value framework, financial mathematics, and monetary architecture for an evolutionary leap into our multi-habitat sustainable future in outer space.

Chapter 11: Introducing Space—introduces the dimension of space, our physical context of matter, into finance. It discusses and refines the layered logic of the conceptualisation of space.

Chapter 12: Respecting Space—introduces the space value of money as a foundational principle necessary to define our relationship with the new dimension and entrench responsibility for impact and respect for space.

Chapter 13: Valuing Space—introduces a set of new equations that could be used to value our investments in space, complimenting the time and risk value of cashflows, integrating space impact.

Chapter 14: Monetising Space—introduces the necessary transformations in our monetary architecture to reflect a space-adjusted financial framework and mathematics.

PART 5: Institutional and Policy Transformations in Space—explores the benefits of a transformed financial value framework, mathematics, and monetary architecture and concludes the book.

Chapter 15: US Debt Ceiling to Wealth Floor—discusses the benefits of the proposed transformations in the context of the US debt ceiling or limit and reveals how it can be transformed into a wealth floor.

Chapter 16: UK New Outer Space Deal—explores the justifications and broad design of a New Outer Space Deal for the UK, to fund and upgrade ambitions, and support its new National Space Strategy.

Chapter 17: Conclusion—summarises the main arguments and concludes the book.

References

Keynes, J. M. 1936. *The General Theory of Employment, Interest and Money*. Macmillan.

Papazian, A. 2023. *Hardwiring Sustainability into Financial Mathematics: Implications for Money Mechanics*. New York: Palgrave Macmillan. https://doi.org/10.1007/978-3-031-45689-3.

Papazian, A. 2022. *The Space Value of Money: Rethinking Finance Beyond Risk and Time*. New York: Palgrave Macmillan. https://doi.org/10.1057/978-1-137-59489-1.

Sagan, C. 1994. *Pale Blue Dot*. Random House.

Part I

Economy, Space, Outer Space, and Human Impact

This part of the book introduces the 'space economy,' as it is commonly referred to, and highlights key features of the sector in absolute and relative terms. It identifies the need for a terminological adjustment and offers a new conceptualisation of space, our physical context of matter, distinguishing between space and outer space. It extends the discussion by providing a snapshot of human impact on space and its many layers, highlighting humanity's evidently brazen disregard and disrespect for space.

2

The New Outer Space Economy

That's one small step for a man, one giant leap for mankind.
Neil Armstrong, Gemini 8 and Apollo 11 Astronaut, 1966

We came all this way to explore the moon, and the most important thing is
that we discovered the Earth.
William Anders, Apollo 8 Astronaut, 1968

Investment banks and large consulting firms have had their eureka moment:
we are in space, and it is good business. Never before have the big firms felt so
inspired by the 'final frontier.' Reports are being published by the who's who
of banking and consulting, and they are all full of positive expectations. The
'space economy,' as they call it, is growing fast (WEF-McKinsey 2024; PwC-
UKSA 2023; Citi 2023; Deloitte 2023; KPMG 2023; Morgan Stanley 2022,
2020).[1] As I elaborate further in the next chapter, I do not use space and
outer space interchangeably, and thus I correct this misnomer by referring to
the *outer space economy*.

While the excitement and positive expectations are grounded in actual
economic factors and data, they do not necessarily imply that investment
bankers and consultants are eagerly waiting for Artemis to take us back to
the moon, or that they are holding their breath for SpaceX's first flight to
Mars. As the discussion in this chapter will reveal, the buzz around the 'space
economy' is not really about outer space exploration and settlement, but

---

[1] This is a small selection of recent publications.

© The Author(s), under exclusive license to Springer Nature
Switzerland AG 2024
A. V. Papazian, *Financing the Race to Space*,
https://doi.org/10.1007/978-3-031-73102-0_2

rather about the growing operationalisation and commercialisation of outer space infrastructure for Earth-based and delivered products and services.

However, it is absolutely true that we are at the dawn of a new era. Very recently, describing orbital launches in 2024, the Space Foundation (2024) states: "averaging a liftoff every 33 hours and 49 minutes, January's 22 successful launches to space marked the busiest start to a year since the Space Age dawned in 1957."

This chapter provides a snapshot of the outer space economy in terms of size and key features. As you will observe, in relative and absolute terms, the new outer space economy is growing, but it is still relatively small. Moreover, it is very much Earthbound, and both the private and public outer space sectors face significant structural constraints.

2.1   Accessing Outer Space: Numbers and Kilograms

To understand the new outer space economy properly, we must begin with the basic facts and figures that describe our access to outer space. It is true that 2024 has ushered in a new era, and January 2024 was the busiest start to a year since 1957 in terms of orbital launches. While we continue to break new ground and build new capabilities, like the SpaceX Starship, we are still at the early stages of the new outer space age. As Table 2.1 and Chart 2.1 reveal, while growing, our access to outer space is still patchy with only a few consistent players.

Table 2.1 and Chart 2.1 provide a six-quarter view, from Q2/2023 to Q3/2024, of orbital launches from Earth. It identifies the number of launches and the countries where the launching entities are headquartered. Thus the case of Rocket Lab is accounted for in the USA numbers and there is no mention of New Zealand as a launch country/site (BryceTech 2024).

The picture revealed through Table 2.1 is interesting. Across the six quarters, from Q2/2023 to Q3/2024, humanity achieved 349 orbital launches. The United States leads the table and accounted for 56.45% of them, followed by China at 28.37%, Russia 6.88%, India 2.29%, Japan 1.72%, France/Europe 1.43%, Iran 1.15, North Korea 1.15%, and South Korea 0.06%. The figures may not sum up to 100 due to rounding. Together, the United States and China account for 84.82% of all orbital launches.

Only three countries have launched every quarter during these six quarters, USA, China, and Russia. India is active across four of the six quarters, with its highest number of launches reaching 3 in Q3/2023. Japan has also achieved 3 launches in a quarter, in Q1/2024. North Korea has launched in four quarters out of the six. Only a few countries have direct access to outer space and the top three are much further ahead.

While the number of orbital launches has been growing steadily, the providers have also been growing and changing. Table 2.2 lists all the launch providers in the six discussed quarters, from Q2/2023 to Q3/2024.

It is truly a remarkable opportunity to witness a global industry, the entire outer space economy of our planet, going through a handful of countries and

Table 2.1 Total orbital launches, and by country, from Q2/2023 to Q3/2024

	Q2/ 2023	Q3/ 2023	Q4/ 2023	Q1/ 2024	Q2/ 2024	Q3/ 2024	TOTAL	Percentage
USA	25	32	30	36	42	32	197	56.45
China	11	20	22	14	16	16	99	28.37
Russia	3	4	6	5	3	3	24	6.88
India	2	3	0	2	0	1	8	2.29
Japan	0	1	0	3	0	2	6	1.72
France/EU	1	1	1	0	0	2	5	1.43
Iran	0	1	0	2	0	1	4	1.15
N Korea	1	1	1	0	1	0	4	1.15
S Korea	1	0	1	0	0	0	2	0.6
Total	44	63	61	62	62	57	349	

Source Compiled by Author from BryceTech (2024)

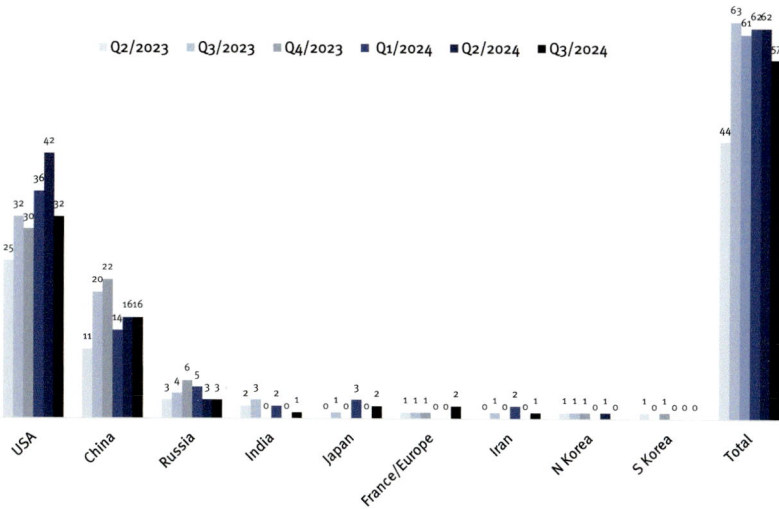

Chart 2.1 Total Orbital Launches and By Country, from Q2/2023 to Q3/2024 (*Source* Compiled by Author from BryceTech 2024)

a small group of providers. The outer space access leagues will undoubtedly and unavoidably change in terms of numbers and composition as we continue to assert our will and develop the technologies of our expansion in outer space.

Table 2.2 lists all the providers identifying the number of spacecraft and the upmass in Kilograms that they have taken to orbit. The providers are first ranked according to the number of spacecraft they launched in Q2/2023. Only six providers have been present across the six quarters; they are: ExPace, United Launch Alliance (ULA), Rocket Lab, Roscosmos, China Aerospace Science & Technology Corp. (CASC), and Space Exploration Technologies (SpaceX). The six providers that have been active across the six quarters, from Q2/2023 to Q4/2024, while not all at the same level, are the consistent element of the broader context where a changing landscape is revealed.

SpaceX and CASC lead the chart. In Q3/2024, SpaceX launched 518 spacecraft with an upmass of 362,087 Kg, and CASC launched 56 spacecraft with an upmass of 30,350 Kg. While the total number of spacecraft and total upmass carried to orbit do not really say much about the technological content and relevance of what is being sent to orbit, they do provide a very accurate picture of outer space access capability. Getting things to orbit is where it all begins after all.

Chart 2.1 and Tables 2.1 and 2.2 describe a very clear picture of our growing but still patchy access to space. The providers, the orbital launches, and their cargo, all together, make up the essential core of the outer space economy. Indeed, the entire outer space economy discussed in the following sections is built upon the launch, and spacecraft and upmass to orbit capabilities of the providers. While I do not discuss or elaborate on here, the list of providers also reflects the geopolitical and military footprint of our access to space.

2.2   The New Outer Space Economy

The outer space economy is growing. Increased activity and output over the last few years have led to positive expectations across the board (Morgan Stanley 2020; Citi 2023; Deloitte 2023; KPMG 2023; PwC-UKSA 2023). The U.S. Bureau of Economic Analysis (BEA 2024a) released a report in June 2024 with updated figures on the US 'space economy' between 2017 and 2022. The data reveals that "the [US] space economy accounted for $131.8 billion, or 0.5%, of total U.S. GDP in 2022. Real GDP grew by 2.3% in the space economy, faster than growth in the overall U.S. economy (1.9%). The

Table 2.2 Number of spacecraft and upmass carried by launch provider from Q2/2023 to Q3/2024

Name of provider	Q2/2023 Number	Kg	Q3/2023 Number	Kg	Q4/2023 Number	Kg	Q1/2024 Number	Kg	Q2/2024 Number	Kg	Q3/2024 Number	Kg
Space One (Canon/IHI)							1	100				
Islamic Revolutionary Guard Corps							1	50			1	34
Orienspace							3	300				
Iranian Space Agency							3	40				
China Manned Space Agency					1	8,100			1	8,082		
National Aerospace Technology Administration (North Korea)*					1	250						
South Korea Ministry of National Defense					1	100						
Mitsubishi Heavy Industries Launch Services			4	3,050			4	4,755			2	175
Chinarocket Co. Ltd.					1	200	9	575			8	4,200

(continued)

Table 2.2 (continued)

Name of provider	Q2/2023		Q3/2023		Q4/2023		Q1/2024		Q2/2024		Q3/2024	
	Number	Kg	Number	Kg	Number	Kg	Number	Kg	Number	Kg	Number	Kg
Galactic Energy			15	1,280	2	265			13	810	7	3,860
Northrop Grumman Space Systems			1	8,051								
Firefly Aerospace			1	200	1	250					8	3,900
Iranian Revolutionary Guards Air Force			1	10								
Landspace			1	1	3	150						
Korean Committee of Space Technology (North Korea)	1	5	1	250					1	100		
China National Space Administration	1	8,082										
Space Pioneer	1	8										
I-Space	1	100										
ExPace	1	20	9	425	8	400	5	350	4	616	4	450
United Launch Alliance (ULA)	1	5,000	3	6,000	2	1,200	7	1,285	2	19,000	3	200
Rocket Lab	1	21	9	416	1	100	10	508	9	285	7	1,212
Roscosmos	2	8,100	4	17,475	7	25,612	24	23,782	13	12,820	4	415
Arianespace	4	5,963	2	6,950	12	806					12	14,760

Name of provider	Q2/2023		Q3/2023		Q4/2023		Q1/2024		Q2/2024		Q3/2024	
	Number	Kg	Number	Kg	Number	Kg	Number	Kg	Number	Kg	Number	Kg
Indian Space Research Organisation (ISRO)	4	2,722	11	5,817			2	2,744			1	60
Korea Aerospace Research Institute (South Korea)	8	202										
CAS Space	26	912					5	890				
GK Launch Services	49	3,168									5	608
China Aerospace Science & Technology Corp. (CASC)	49	23,069	24	24,560	31	40,810	27	29,426	23	35,534	56	30,350
Space Exploration Technologies (SpaceX)	648	214,095	519	381,278	590	382,080	525	429,125	659	530,488	518	362,087
Total	**797**	**271,467**	**605**	**455,763**	**662**	**460,573**	**626**	**483,930**	**725**	**607,735**	**639**	**422,569**

Source Compiled by Author from BryceTech (2024)
* National Aerospace Technology Administration is the official space agency of the Democratic People's Republic of Korea, North Korea. It succeeded the Korean Committee of Space Technology. The use of both names follows the Bryce Briefings verbatim (BryceTech 2024).

statistics also show in 2022 the space economy accounted for $232.1 billion of gross output."[2]

In April 2024 the World Economic Forum and McKinsey & Company (WEF-McKinsey 2024) released a report titled 'Space: The $1.8 Trillion Opportunity for Global Economic Growth.' According to WEF-McKinsey (2024) the global 'space economy' stood at $630 billion in 2023 and is expected to reach $1.8 trillion by 2035, growing at 9% per year.

> The space economy is forecast to soar to $1.8 trillion by 2035 in an increasingly connected and mobile world, impacting and creating value for nearly all industries on Earth and providing solutions to many of the world's greatest challenges…. The space economy is forecast to reach $1.8 trillion by 2035, up from $630 billion in 2023 and growing at an average of 9% per annum – well above the growth rate of global gross domestic product (GDP). (WEF-McKinsey 2024, 4)

If we were to compare the $630 billion figure to global GDP in 2023, we observe, as in Chart 2.2 and 2.3, that it represents 0.6% of global GDP at $104,476.43 billion in 2023 (IMF 2024). In the same year, global military expenditure at $2443.4 billion amounted to 2.3% of global GDP (SIPRI 2024).

Looking at the global wealth figures (Table 2.3) published by UBS in the Global Wealth Report (UBS 2023), the 2023 outer space economy at $630 billion represents 0.13% of global wealth in 2022, at $454,385 billion. Wealth is defined as the value of financial assets plus real assets owned by households, minus their debts.

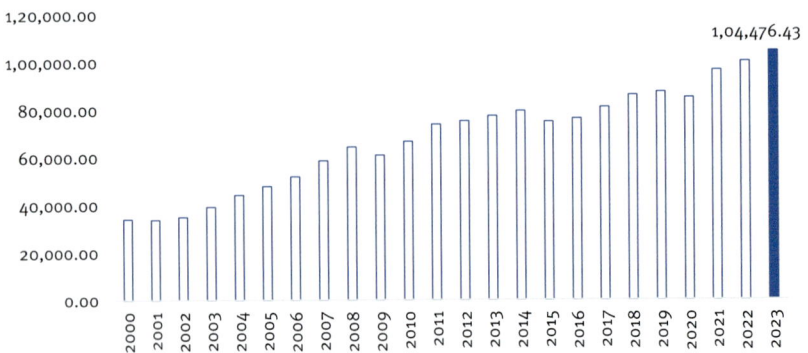

Chart 2.2 Global GDP in Current US Dollars, in billions (*Source* IMF [2024])

---

[2] Note that Gross Output (GO) is a measure of an industry's sales receipts, including sales to final users in the economy (GDP) and sales to other industries, i.e., intermediate inputs (BEA 2024b).

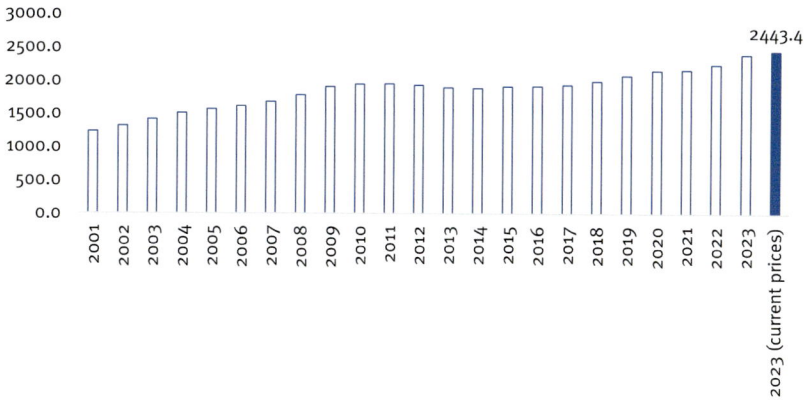

Chart 2.3 World Military Spending, from 2001 to 2023, in US$ billions, 2022 prices & rates (*Source* SIPRI [2024])

Interestingly, again in 2023, the top 15 oil and gas companies around the world achieved revenues of $2857.19 billion, which is around 454% of the entire outer space economy (Statista 2024a). As Chart 2.4 depicts, this exciting new outer space economy is also smaller than the revenues of the largest American and largest Chinese oil and gas companies (Sinopec and ExxonMobil), together at $686.42 billion.

Obviously, the oil and gas sector, which provides energy to a variety of others, is not a like for like comparison. The purpose here is to contextualise the outer space sector. Interestingly, despite past and expected growth rates, and the $630 billion figure for 2023, the outer space sector was also smaller than global advertising spending at $1,012.78 billion, which is expected to rise to 1,076.47 billion in 2024 (Statista 2024b) (Chart 2.5).

When considering outer space economy figures the first most important factor to consider are the regulations and controls that directly affect the sector. Due to the sensitive nature of the technologies involved, they are often under strict export controls. Taking the example of the United States, which is the largest player in the outer space economy in dollar terms, there are a number of regulatory frameworks and departments that constrain the size and growth of the sector.

The International Traffic in Arms Regulations (ITAR) process has been developed under the jurisdiction of the Department of State and is administered by the Directorate of Defense Trade Controls (DDTC). The ITAR process controls items, information, or activities that could be used for threatening foreign military purposes, whether actual products (defense articles) or assistance (defense services).... The Export Administration Regulations (EAR)

Table 2.3 2022 Global wealth key features

	Total wealth	Change in total wealth		Wealth per adult		Change in financial assets		Change in non-financial assets		Change in debt	
	USD bn	USD bn	%	USD bn	%	USD bn	%	USD bn	%	USD bn	%
Africa	5,909	85	1.5	8,345	-1.3	57	2.1	36	1	8	1.9
Asia–Pacific	77,974	-2,070	-2.6	61,154	-4.0	-2,931	-6.5	476	1	-385	-3.6
China	84,485	-1,462	-1.7	75,731	-2.2	-116	-0.3	-1,632	-3.1	-285	-2.8
Europe	104,410	-3,703	-3.4	177,179	-3.4	-5,736	-10.4	1,552	2.3	-480	-3.2
India	15,365	675	4.6	16,500	2.8	34	1	679	5.4	38	3
Latin America	15,071	2,359	18.6	32,760	16.9	819	12.9	1,745	22.7	204	15.1
North America	151,170	-71,766	-4.5	531,826	-5.3	-11,226	-9.0	5,025	9.5	965	4.9
World	454,385	-11,281	-2.4	84,718	-3.6	-19,099	-6.8	7,882	3.2	65	0.1

Source UBS (2023)

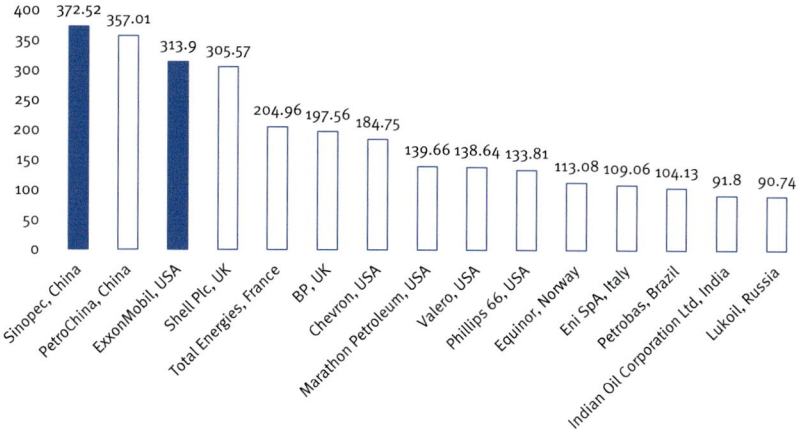

Chart 2.4 Leading oil and gas companies worldwide based on revenue as of 2023 (*Source* Statista [2024a])

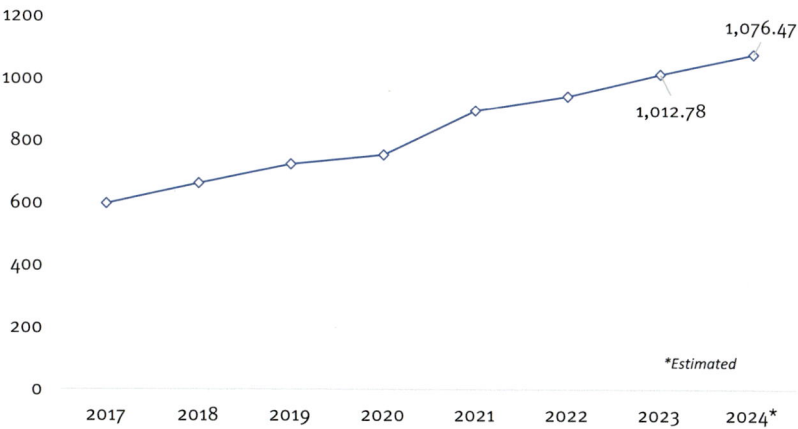

Chart 2.5 Global Advertising Spending 2017–2024 (*Source* Statista [2024b])

process controls items and technologies considered to be "dual use", meaning applicable to commercial or military use. These items are detailed in the EAR under the Commerce Control List (CCL). The Commerce Department's Bureau of Industry and Security (BIS) administers the EAR process.... The vast majority of commercial spacecraft and components fall under the jurisdiction of the EAR. (OSC 2017, 5)

The second most important factor relevant to our understanding of the outer space economy is the fact that the sector is quite broadly defined. The US Bureau of Economic Analysis defines it as follows (BEA 2020): "[t]he space economy consists of space-related goods and services, both public and private.

This includes goods and services that: are used in space, or directly support those used in space; require direct input from space to function, or directly support those that do; are associated with studying space."

The Organisation for Economic Cooperation and Development (OECD 2022) identifies the following main sectors of 'space activities:' satellite communications; positioning, navigation and timing; Earth observation; space transportation; science; space technologies, and generic technologies or components that may enable space capabilities.

Within the diverse landscape of the growing new 'space economy,' Morgan Stanley (2022) identifies five main themes: a growing relationship between space and climate change, increased capital formation, mitigating orbital debris, space and security, and finally telecoms (Morgan Staley 2022). Indeed, in its 2023 report, the Satellite Industry Association (SIA 2023) reveals that in 2022 the private space economy generated revenues of $384 billion and the commercial satellite industry was responsible for $281 billion, 73% of the total business revenues. According to SIA's latest report (2024), in 2023 "a historic number of launches deployed a record number of satellites" and "satellite manufacturing, ground equipment and launch revenues all continued to increase in 2023".

The WEF-McKinsey (2024) report breaks down the 'space economy' into two core parts, backbone and reach.

> What is the space economy? If you're picturing satellites, launchers and services like broadcast television (TV) or the global positioning system (GPS), you might only be thinking of half of it. These "backbone" applications – those where revenues accumulate directly to space hardware and service providers – made up slightly greater than 50% of the global space economy in 2023, or $330 billion.
>
> Meanwhile, space is playing a key role in enabling companies across industries to generate revenues. Entire markets would not exist but for these space technologies, which are considered the "reach" of the space economy. For example, without the combination of satellite signals and chips inside your smartphone, Uber would never have reached such a global scale with its unique ability to connect drivers and riders and provide directions in every city. (WEF-McKinsey 2024, 9)

Table 2.4 reveals in more detail the functional breakdown of the global market considering the backbone and the reach applications within the sector. Note that some figures differ across sources due to methodological differences.

Looking at Table 2.4 we observe that out of all the applications, including commercial and state-sponsored, civil and defence, infrastructure and services, amounting to $630 billion in 2023, space exploration and space observation together amounted to $18 billion, which represents 2.86% of the entire outer space economy. Interestingly, even though both applications are set to grow to $44 billion by 2035, their percentage of the total is set to decrease to 2.46%.

Naturally, there are other applications that should also be considered relevant. Including other categories like launch systems and operations, launch sites and operations, ground hardware and operations, basic sciences, technology and research, and others (commercial and civil) improves the numbers but does not change the overall picture. Outer space exploration is a relatively small part of the outer space economy.

Looking at broader industry contributions to the 'space economy,' the WEF-McKinsey (2024) report reveals that the 'food and beverage,' 'supply chain and transportation,' 'retail, consumer goods, and lifestyle,' and 'media, entertainment, and sports' industries together make up 61.4% of the outer space economy in 2023 at $387 billion, and $1073 billion or 59.9% of the projected outer space economy in 2035 (see Table 2.5).

This representation of the 'space economy' is a true description of the fact that many of the applications and activities of this growing sector are Earthbound.[3] Only a small fraction of them is about outer space exploration and settlement. There is a simple reason for this. All private outer space industry players have to sell their products and services to survive, and all the customers and the money supply they must pursue are on Earth. In other words, a growing outer space economy implies growing industries that provide services that use outer space for Earth-based services and products. Indeed, only a few companies aim beyond Earthbound services.

I will expand and discuss the Earthbound nature of the current outer space economy in the following chapters, specifically in relation to our financial value framework, mathematics, and monetary architecture. For now, let me restate the main point. The challenge of the private outer space economy is to go from 'A' to 'B' while collecting 'C' that can only be found in 'A.' The cash flow imperative imposes very Earth-centric priorities on the private

---

[3] It is quite fascinating to note that the 'space economy' described and discussed by WEF-McKinsey (2024) seems to omit the solar energy sector. Given the definition of the 'space economy' provided by US Bureau of Economic Analysis discussed earlier, it appears that 'require direct input from space to function, or directly support those that do' does not actually apply to celestial bodies, but only to human technology. Moreover, the WEF-McKinsey (2024) discussion of the 'space economy' bypasses orbital launches and their providers, even though they represent the very infrastructure the 'space economy' relies on.

Table 2.4 Global 'space economy' market sizing ($ billion in 2023–$ billion in 2035)

Global space economy $630 billion–$1790 billion

Category	Segment	2023	2025	Reach	2023	2025
Backbone		330	755	Reach	300	1035
Commercial 205–435						
Services and end user equipment 183–356	Communications	133	218	Enabled by communications	67	133
	PNT	47	120	Enabled by PNT	196	799
	Space-based Earth observation (EO)	2	9	Enabled by EO	2	10
	Others[1]	1	9			
Infrastructure and Support 22–79	Vehicles and satellite manufacturing	17	44	Enabled by others	4	16
	Launch sites and operations	2	3			
	Ground hardware and operations	2	11			
	Others[2]	1	21			
State-sponsored 125–320	Communications	4	8	Enabled by civil	3	6
Civil 59–140	Basic sciences, technology and research	8	12			
	Space-based Earth observation	14	32			
	Space exploration	15	39			
	Space observation	3	5			
	Launch systems and operations	5	11			

Global space economy $630 billion–$1790 billion

			Enabled by defence	28	71
Defence 66–180	Others³			10	24
	Command, control and communications	4	16		
	Space sensing	7	22		
	PNT	4	10		
	Space domain awareness and com bat pow	1	4		
	Launch system s and operations	5	12		
	Classified defence and intelligence	37	91		
	Others³	8	25		

[1]E.g. space tourism (aside from launch services), mining, in-space manufacturing
[2]E.g. in-orbit servicing and de-orbiting, insurance for space systems, comm ercial participation for space stations & lunar missions
[3]Miscellaneous administrative and research costs
PNT: Positioning, Navigation and Timing
EO: Earth Observation
Source Adapted by author from WEF-McKinsey (2024)

outer space sector, even for those companies who do have exploration and settlement objectives.

Taking the unique example of SpaceX, the company clearly states on its mission page that it intends to make humanity a multiplanetary species, something that founder Elon Musk has written and spoken about numerous times (Musk 2017). While the vision and end goal may be the colonisation of Mars, short-term cash flow generating activities are necessary for the company's survival and growth. This explains many of the company's activities like flying astronauts to the International Space Station, selling seats for private travellers, providing Starlink internet, taking cargo to orbit for other customers, etc.

Upon the release of the Q1 2024 launch report (BryceTech 2024), which revealed that SpaceX was leading everyone in terms of spacecraft launches as well as upmass carried to orbit (See Charts 2.6 and 2.7), on the 20th of May 2024, replying to a tweet on X, Elon Musk wrote: "SpaceX might exceed 90% of all Earth payload to orbit later this year. Once Starship is launching at high rate, probably > 99%. Has to be or we can't build a city on Mars or

Table 2.5 'Space economy' by industry ($bn USD)

		2023	2035
1	Supply chain and transportation	88	412
2	Food and beverage	100	334
3	Media, entertainment, and sports	143	157
4	Retail, consumer goods and lifestyle	56	170
5	State-sponsored – defence	94	251
6	State-sponsored – civil	62	146
7	Digital communications	19	70
8	Space	22	67
9	Aviation and aerospace - non-space	14	34
10	Agriculture	5	33
11	Information technology	7	25
12	Engineering and construction	7	21
13	Professional services	5	17
14	Automotive and manufacturing	6	17
15	Insurance and asset management	1	13
16	Energy	1	7
17	Banking and capital markets	< 1	7
18	Travel and tourism	< 1	6
19	Global health and healthcare	< 1	2
20	Mining and metals	< 1	1
	Total	**630**	**1,790**

Source WEF-McKinsey (2024)

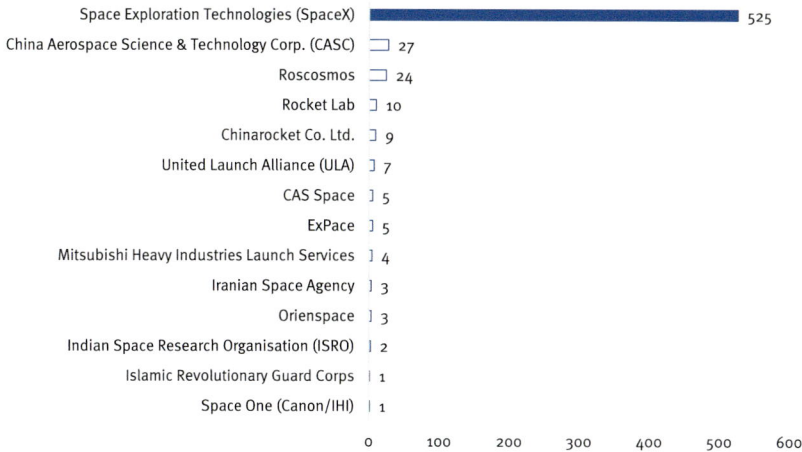

Chart 2.6 Number of Spacecraft Launched by Provider Q1/2024 (*Source* BryceTech [2024])

base on moon. We file almost no patents, so nothing stopping competition from copying us" (Musk 2024a). While tweets should always be taken with a few grains of salt, this specific statement by Elon reflects the above-described intention and necessity. On the 22nd of September 2024, Elon Musk tweeted again saying: "By the way, our commercial Starlink program is the primary source of funding for Starship (NASA is helping too). So thank [you] for buying Starlink and supporting humanity's future in space. If you look closely at your Starlink router, you will notice that it has an illustration of the Earth-Mars transfer orbit" (Musk, 2024b).

This is why *public outer space* expenditures are critical for any meaningful and sustainable development in outer space. Public expenditures on outer space agencies and programs do not expect a monetary return, and they are not constrained by the above imperative. Public budgets, however, have different kinds of constraints. For now, let's look at the figures.

2.2.1 Public Outer Space Sector Expenditures

In 2023 total global governmental expenditure on outer space development programs and agencies reached $117 billion (Euroconsult 2023). Compared to the entire outer space economy size measured by WEF-McKinsey (2024), this $117 billion figure represents 18.6%. In other words, in 2023, global public outer space expenditures amounted to 0.1% of world GDP (IMF 2024) and 0.03% of global wealth (UBS 2023). Meanwhile, in the same year, total global military expenditure reached $2443 billion (SIPRI 2024),

Provider	Upmass (Kg)
Space Exploration Technologies (SpaceX)	4,29,125
China Aerospace Science & Technology Corp. (CASC)	29,426
Roscosmos	23,782
Mitsubishi Heavy Industries Launch Services	4,755
Indian Space Research Organisation (ISRO)	2,744
United Launch Alliance (ULA)	1,285
CAS Space	890
Chinarocket Co. Ltd.	575
Rocket Lab	508
ExPace	350
Orienspace	300
Space One (Canon/IHI)	100
Islamic Revolutionary Guard Corps	50
Iranian Space Agency	40

0   1,00,000   2,00,000   3,00,000   4,00,000   5,00,000

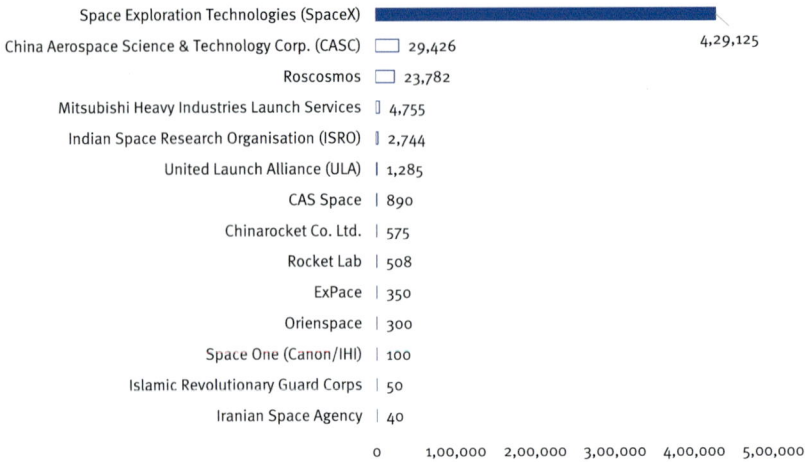

Chart 2.7 Upmass Carried by Provider in Kg Q1/2024 (*Source* BryceTech [2024])

representing 2.3% of world GDP. Military spending on outer space programs has also risen sharply and overtaken civilian spending.

The militarisation of public outer space spending is led by the United States (Euroconsult 2023; Breaking Defense 2023; Space Foundation 2023), mainly since the inception of the US Space Force in 2019 during President Donald J. Trump's first term.[4],[5] Between NASA, the US Space Force (USSF), and intelligence spending, the US allocated $73 billion to space activities in 2023, roughly 63% of the total $117 billion (Euroconsult 2023). Charts 2.8 and 2.9 depict the NASA and USSF budgets. The militarisation of outer space and the growth in military spending is one of the key features of public governmental budgets and expenditures on outer space.

The other overarching feature of public outer space expenditures is that they are subject to governmental budget cuts due to fiscal and monetary conditions, government revenues, and debts. In 2024, describing recent budget cuts in an article titled 'NASA's budget woes put ambitious space research at risk,' Mann (2024) writes:

> Dreams of exploring the cosmos have crashed up against the harsh reality of budget cuts in the United States. Congressional approval of the 2024 federal budget earlier this year left NASA with roughly half a billion dollars less than

---

[4] "The U.S. Space Force was established Dec. 20, 2019 when the National Defense Authorization Act was signed into law (with bi-partisan support), creating the first new branch of the armed services in 73 years" (USSF 2024).

[5] Goswami and Garretson (2020) provide a detailed account of 'the great power competition to control the resources of outer space.'

the agency had in 2023 — and Mars science has taken the biggest hit. (Mann 2024)

Discussing NASA's recent budget requests and appropriations, the Planetary Society writes:

> The President's Budget Request for NASA for fiscal year 2025 is $25.4 billion, a 2% increase over 2024 and the same as 2023. The budget proposes flat or modest cuts to most directorates within the agency, a noted departure from previous plans, which originally called for billions of additional dollars over the coming years…This discrepancy is a consequence of strict spending caps passed by Congress in 2023, which functionally froze U.S. discretionary spending for two years. These arbitrary caps create a zero-sum game between federal agencies, and NASA, while respected and admired amongst lawmakers, is rarely a top priority. (Planetary Society 2024b)

The strict spending caps passed by the US Congress and the fact that space exploration is 'rarely a top priority' reveal the constraints on the public outer space sector. While not obliged to pursue and chase earthly money supply, public budgets impose a unique set of limitations on our ability to invest and expand in outer space. Similarly, announced in 2024, the US Space Force's 2025 budget registered its first ever decrease, albeit a small one (Charts. 2.8, 2.9).

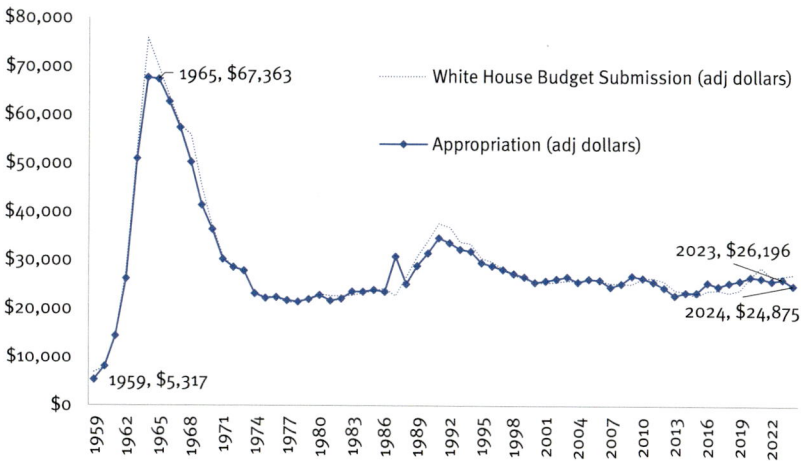

Chart 2.8 NASA Budget – Adjusted for Inflation ($mn USD) 1959–2024 (*Source* Planetary Society [2024a])

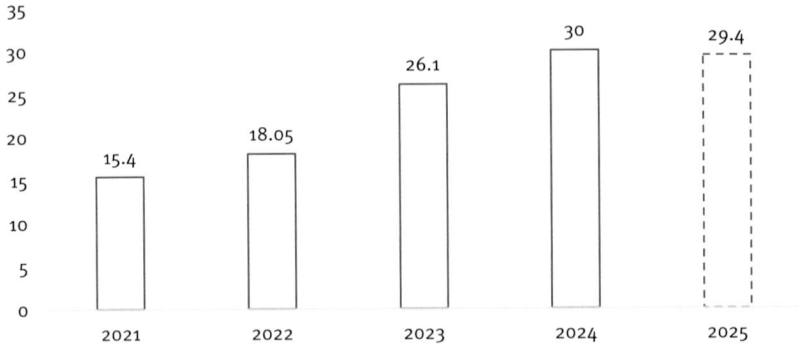

Chart 2.9 US Space Force Budget ($bn USD) 2021–2025 (*Source* USAF-FMC [2024])

Table 2.6 presents the US Federal Government budget for 2023, including revenues, outlays, and the deficit. The purpose here is to identify the dependencies of public outer space expenditure in the US, which falls in the discretionary outlays of the federal government. While the private sector is chasing earthly money supply, the public outer space sector is constrained by the political and economic frameworks that shape and define public funding and spending.

In its recently released report the U.S. Bureau of Economic Analysis (BEA 2024a) lists all the Federal Government agencies and funded research and development centres that are involved in 'space activities.' Table 2.7 is a reproduction that reveals the wide range of public institutions involved in the public side of the US 'space economy.' Once again, we can see clearly that the composition in the public side is also diverse, indicating that not all have outer space exploration, development, and settlement as a top priority.

2.3   Funding and Investment Flows in the Outer Space Economy

To sum up the previous discussion, according to the latest measurements and projections by WEF-McKinsey (2024) the outer space economy stood at $630 billion in 2023 and is projected to reach $1.8 trillion by 2035. At $630 billion, the new outer space economy represents 0.6% of world 2023 GDP at $104,476.43 billion (IMF 2024), and 0.13% of 2022 global wealth at $454,385 billion (UBS 2023). In comparison, global military expenditure in 2023 at $2443 billion amounted to 2.3% of global GDP (SIPRI 2024). Again in 2023, the top 15 oil and gas companies around the world had revenues of $2857.19 billion, which is around 454% of the entire outer space

Table 2.6 US federal government revenues, outlays and deficit, 2023, in Trillions

OUTLAYS	Dollars	% of GDP
Total Outlays	**6.086**	**22.70%**
Mandatory	*3.705*	*13.90%*
Social Security	1.3	5.00%
Medicare	0.839	3.10%
Medicaid	0.616	2.30%
Income Security Programs	0.448	1.70%
Other	0.502	1.90%
Discretionary	*1.722*	*6.40%*
Nondefense	0.917	3.40%
Defense	0.805	3.00%
Net Interest on Debt	*0.659*	*2.40%*
REVENUES	Dollars	% of GDP
Total Revenues	**4.4**	**16.50%**
Individual Income Taxes	2.2	8.10%
Payroll taxes	1.6	6.00%
Corporate Income taxes	0.42	1.60%
Other	0.229	0.80%
DEFICIT	Dollars	% of GDP
Deficit	**1.7**	**6.30%**
Average Deficit over the past 50 years		3.70%
Debt held by the public at end of 2023		97%
Change in net interest costs from 2021 to 2023		87%

Source CBO (2024)

economy (Statista 2024), and the revenues of two oil and gas companies, the largest American and the largest Chinese company, at $686.42 billion, were larger than the entire outer space economy. We also observed that the outer space economy at $630 billion was smaller than global advertising spending in 2023, at $1,012.78 billion. Looking at the industry composition of the sector, according to the WEF-McKinsey report (2024) again, the 'food and beverage,' 'supply chain and transportation,' 'retail, consumer goods, and lifestyle,' and 'media, entertainment, and sports' industries together make up 61.4% of the outer space economy in 2023.

Meanwhile, global governmental expenditure on outer space programs and agencies amounted to $117 billion in 2023, representing 0.1% of world GDP and 0.03% of global wealth and slightly surpassing the revenues of the global toilet paper market at $107.38 billion (Euroconsult 2023; Statista 2023a). The United States accounts for 63% of total global government expenditure on space agencies and programs, also leading in military spending in the sector. The NASA budget has been relatively stagnant over the years

Table 2.7 Federal government agencies and federally funded research and development centres with direct space activity

Agency or Centre	
Nondefense	
U.S. Department of Commerce	National Institute of Standards and Technology
U.S. Department of Energy	National Oceanic and Atmospheric Administration
U.S. Department of the Interior	National Science Foundation
U.S. Department of Transportation	Smithsonian Institution
NASA (National Aeronautics and Space Administration)	
Defence	
U.S. Air Force	Missile Defense Agency
U.S. Army	U.S. Navy
Defense Advanced Research Projects Agency	Office of the Secretary of Defense
Defense Information Systems Agency	Space Development Agency
Defense Innovation Unit	U.S. Space Force
Defense Logistics Agency	
Federally funded research and development centres	
Aerospace	Los Alamos National Laboratory
Brookhaven National Laboratory	National Center for Atmospheric Research
Jet Propulsion Laboratory	National Optical Astronomy Observatory
Lawrence Livermore National Laboratory	National Radio Astronomy Observatory
Lincoln Laboratory	National Solar Observatory

Source BEA (2024a)

and declined in 2024 due to budget cuts. The US Space Force 2025 budget registered a decline, for the first time since inception.

In other words, the species spends more on trying to sell goods and services to itself than on building its collective future in outer space; the species spends more on trying to 'defend' itself from itself than on building its collective future in outer space. Moreover, the species allocates 0.6% of world GDP and 0.13% of its wealth to the outer space economy and the large part of that allocation is for Earth-based products and services. Last but not least, the species allocates to the public space economy 0.1% of world GDP and 0.03% of global wealth.

The private outer space economy is primarily Earthbound given that it must sell its products and services on Earth, where the money supply and customers are. Thus, its priorities are by default Earthbound, and outer space development, exploration, and settlement are secondary to this primary reason for being. Indeed, the outer space economy encompasses a vast array of industries that *use* outer space, and only a small proportion is dedicated to actual outer space exploration and settlement.[6]

Meanwhile, public outer space expenditures are free from such constraints as they do not pursue any monetary return when setting priorities and spending budgets. However, other types of limitations apply to public sector investments. They are constrained by governmental budgets, tax revenues, public debts, and budget cuts. Echoing the same diversity, specifically in the US, a diverse group of departments and institutions are involved in 'space activities,' and they are not all focused on achieving outer space exploration and settlement.

These limitations are closely linked to and defined by our financial value framework, mathematics, and monetary architecture. I explore them in detail in later chapters. This link and influence are born out of the funding sources and investment flows in the outer space sector.

Figure 2.1 provides a high-level overview of funding and investment sources and flows into the outer space economy, revealing that there is also an overlap between the two when private companies are subcontracted by public agencies to deliver products and services. Indeed, on the 26th of June 2024 NASA awarded SpaceX a $843 million contract to bring the International Space Station (ISS) out of orbit at the end of its lifespan in 2030. This is just one example of course. In 2014, NASA awarded Boeing and SpaceX multibillion contracts for commercial crew transportation systems (Foust 2014).[7]

While the private outer space sector funds its investments through equity, debts, and revenues, the public outer space sector funds its investments or expenditures through taxes and debts. Naturally, the private sector includes many different kinds of firms operating across the sectoral and application landscape discussed in the previous section. Aerospace-focused examples in each category would be Boeing (Public Company Investments) listed on the New York Stock Exchange (NYSE), SpaceX (Private Company Investments), Firefly Aerospace (Private Startup Investments).

I have included non-profit organisations and foundations separately in Fig. 2.1 in order to highlight the fact that, while monetarily not as significant in comparison, they have always played an important role in the outer space

---

[6] See EIB (2019) for a detailed value chain mapping of the industry.

[7] In most of the reports looking at the outer space economy, this overlap between private revenues and public expenditures is not always addressed. As such, the figures may vary once such overlaps are taken into account.

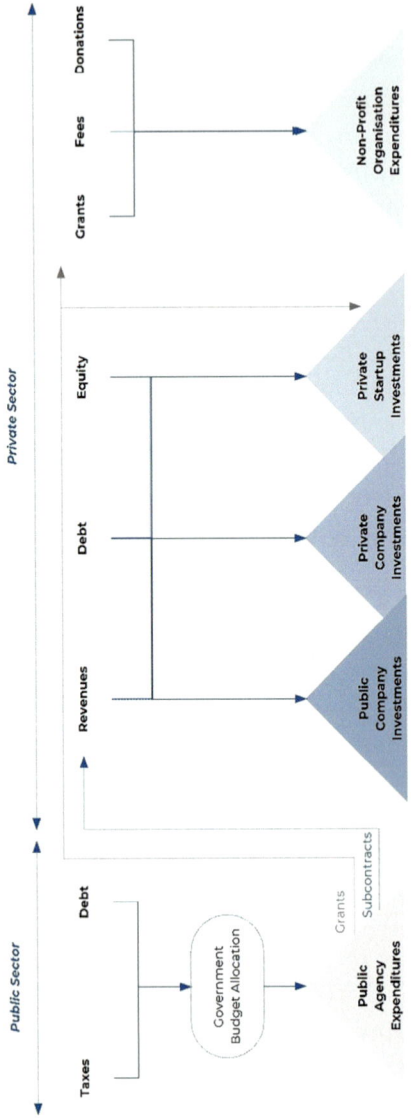

Fig. 2.1 Outer space economy main funding sources and investment flows (*Source Author*)

sector. Whatever their registration status, whether membership organisations, charities, or unincorporated initiatives, non-profits have always had their own unique contribution and have often shaped the debate in the outer space sector. With different degrees of sophistication, non-profits have also defied our existing financial and monetary framework to contribute and guide our evolution in outer space. Funded through grants, donations, and membership fees, the non-profit sector has had its unique impact.

The British Interplanetary Society, the International Astronautical Federation, the Planetary Society, the National Space Society, the Space Foundation, the Space Science Institute and Space for Humanity are just a few well known examples, but the list is long and global. Indeed, this very book has been possible thanks to the support of the Space Value Foundation. While the dedication and contribution of the non-profit niche in the outer space sector is significant, it is primarily focused on research, education, and advocacy.

2.4   Conclusion

If humanity is going to be able to invent, manufacture, deploy, and maintain lunar and Mars habitats and other exploration missions beyond LEO, MEO, and GEO[8], it must be able to *finance and sustain* massive investment programs aimed at building, deploying, and maintaining the off-Earth structures. It must also invest to prepare the needed on-Earth and off-Earth workforces. Such ventures require an entirely new level of monetary resources. Moreover, if the outer space sector, private and public, is continuously chasing customers and cash flows to survive and is consistently undermined by budget cuts, it will be impossible to truly extend our reach, even if we manage to get to Mars. This dependence on Earthly money supply and limited debt-funded government budgets is a restriction when it comes to extending and maintaining our footprint in outer space. This is especially true given the great distances and time horizons involved. I discuss these and other relevant aspects in later chapters.

The current outer space economy, however positive and encouraging its momentum, is Earthbound. The vast majority of companies and projects involved are in sectors who have no outer space ambitions. They utilise outer space infrastructure, such as satellite communications, to deliver their products and services on Earth. SpaceX is a very rare and unique phenomenon, most definitely thanks to the vision and character of Elon Musk himself. As

---

[8] Low Earth Orbit, Medium Earth Orbit, Geostationary Orbit.

he often states and clearly expresses, his ultimate goal is to make humanity a multiplanetary species. In fact, as per his published articles, interviews, company pages, and tweets, he wants to grow and expand the cash flow generating products and services through access to space in order to be able to build the habitats of the future. Naturally, there are technical and knowledge benefits to the parallel services SpaceX offers, but they are hardly the only way to acquire such knowledge and experience. As such, one wonders what we could achieve, and how fast, if SpaceX had funding to simply focus on our outer space exploration and settlement vision.

In a 2005 interview, former NASA Administrator Michael D. Griffin stated that "[t]he goal isn't just scientific exploration. … It's also about extending the range of human habitat out from Earth into the solar system as we go forward in time. … In the long run a single-planet species will not survive. … If we humans want to survive for hundreds of thousands or millions of years, we must ultimately populate other planets. Now, today the technology is such that this is barely conceivable. We're in the infancy of it" (Washington Post 2005).

While breakthroughs are not guaranteed, and the journey will naturally take time, how much we achieve, and how quickly, depends on the resources we allocate to the above articulated vision and mission. NASA's budget and achievements in the last 66 years substantiate this. In the seven years preceding the moon landing, from 1963 to 1969 inclusive, NASA's inflation adjusted budget amounted to $420.78 billion, on average $60 billion a year; in the seven years between 2017 and 2023 inclusive, NASA's budget amounted to $180.1 billion, on average $25.7 billion per year (Planetary Society, 2024a).

Between the budgetary constraints imposed on public space agencies and the cash flow imperative of private firms, we are, in truth, chained to the surface of this planet. We must be able to allocate more funding to the outer space exploration, development, and settlement endeavour. In the right hands and supporting the right private and public entities, additional funding can free the key front runners and ground-breaking firms to achieve more, faster, and better—and without any artificial constraints. Indeed, the above limitations are artificially imposed on us by our own interpretations and assumptions.

How we can increase the funding levels of the outer space sector and why we seem unable to do so now is the main theme of this book. Indeed, all of the above discussed funding/investment flows into the sector are directly and indirectly defined by our financial value framework, financial mathematics, and monetary architecture, where, as I argue in the following discussion, our current limitations are born.

References

BEA. 2024a. New and Revised Statistics for the U.S. Space Economy, 2017– 2022. U.S. Bureau of Economic Analysis. https://apps.bea.gov/scb/issues/2024/ 06-june/0624-space-economy.htm. Accessed 29 June 2024.

BEA. 2024b. Gross Output by Industry. U.S. Bureau of Economic Analysis. https:// www.bea.gov/data/industries/gross-output-by-industry. Accessed 29 June 2024.

BEA. 2020. Preliminary Estimates of the U.S. Space Economy, 2012–2018. Bureau of Economic Analysis. https://apps.bea.gov/scb/issues/2020/12-december/1220- space-economy.htm. Accessed 12 March 2024.

Breaking Defense. 2023. Led by US, Global Spending on Military Space Jumped to $54B in 2022: Space Foundation. https://breakingdefense.com/2023/07/led- by-us-global-spending-on-military-space-jumped-to-54b-in-2022-space-founda tion/. Accessed 12 April 2024.

BryceTech. 2024. Brycetech Briefing. Global Orbital Space Launches Q2/2023 to Q3/2024. https://brycetech.com/briefing. Accessed 09 December 2024.

CBO. 2024. The Federal Budget in Fiscal Year 2023. Congressional Budget Office. https://www.cbo.gov/system/files/2024-03/59727-Federal-Budget.pdf. Accessed 12 June 2024.

Citi. 2023. Space: High Power, Citi Global Insights. https://www.citigroup.com/glo bal/insights/global-insights/space-high-power-. Accessed 12 December 2023.

Deloitte. 2023. Riding the Exponential Growth in Space. Deloitte Insights. https:// www2.deloitte.com/us/en/insights/industry/aerospace-defense/future-of-space- economy.html. Accessed 12 December 2023.

EIB. 2019. The Future of the European Space Sector: How to Leverage Europe's Technological Leadership and Boost Investments for Space Ventures. https:// www.eib.org/attachments/thematic/future_of_european_space_sector_en.pdf. Accessed 12 December 2023.

Euroconsult. 2023. New Historic High for Government Space Spending Mostly Driven by Defense Expenditures. https://www.euroconsult-ec.com/press-release/ new-historic-high-for-government-space-spending-mostly-driven-by-defense-exp enditures/. Accessed 12 January 2024.

Foust, J. 2014. NASA Selects Boeing and SpaceX for Commercial Crew Contracts. Space News. https://spacenews.com/41891nasa-selects-boeing-and- spacex-for-commercial-crew-contracts/. Accessed 12 April 2024.

Goswami, N., Garretson, P. 2020. Scramble for the Skies: The Great Power Competi- tion to Control the Resources of Outer Space. Lexington Books.

IMF. 2023a. Fossil Fuel Subsidies Surged to Record $7 Trillion. IMF Blog. https://www.imf.org/en/Blogs/Articles/2023/08/24/fossil-fuel-subsidies-sur ged-to-record-7-trillion. Accessed 12 December 2023.

IMF. 2024. Global GDP in Current US Dollars. International Monetary Fund. https://www.imf.org/external/datamapper/NGDPD@WEO/OEMDC/ADVEC/WEOWORLD. Accessed 12 March 2024.

KPMG, 2023. A Prosperous Future: Space. KPMG and AMCham. https://assets.kpmg.com/content/dam/kpmg/au/pdf/2023/prosperous-future-report-space.pdf.Accessed 12 January 2024.

Mann, A. 2024. NASA's budget woes put ambitious space research at risk. Science News. https://www.sciencenews.org/article/nasas-budget-woes-space-research-risk. Accessed 9 May 2024.

Morgan Stanley. 2022. 5 Key Themes in the New Space Economy. https://www.morganstanley.com/ideas/space-economy-investment-themes. Accessed 12 May 2024.

Morgan Stanley. 2020. Space: Investing in the Final Frontier. https://www.morganstanley.com/ideas/investing-in-space. Accessed 12 January 2024.

Musk, E. 2017. Making Life Multiplanetary. SpaceX. https://www.spacex.com/media/making_life_multiplanetary_transcript_2017.pdf. Accessed 12 March 2024.

Musk, E. 2024a. Tweet on Q1 2024 Orbital Launch Report. https://x.com/elonmusk/status/1792689139470704718. Accessed 20 May 2024.

Musk. E. 2024b. Tweet on Starship Funding. https://x.com/elonmusk/status/1837913942859067438?s=12&t=BF3HJcIF09VqnoOyfOPkgA.Accessed 01 October 2024.

Musk, E. Making Life Multiplanetary. SpaceX. https://www.spacex.com/media/making_life_multiplanetary_transcript_2017.pdf. Accessed 12 March 2024.

OECD. 2022. OECD Handbook on Measuring the Space Economy, 2nd Edition. The Organization for Economic Cooperation and Development. https://www.oecd-ilibrary.org/science-and-technology/oecd-handbook-on-measuring-the-space-economy-2nd-edition_8bfef437-en. Accessed 02 February 2024.

OSC. 2017. U.S. Export Controls for the Commercial Space Industry. Office of Space Commerce. Department of Commerce and Federal Aviation Administration. https://www.space.commerce.gov/wp-content/uploads/2017-export-controls-guidebook.pdf. Accessed 12 June 2024.

Planetary Society. 2024a. Your Guide to NASA's Budget. Planetary Society. https://www.planetary.org/space-policy/nasa-budget. Accessed 12 April 2024.

Planetary Society. 2024b. NASA's FY 2025 Budget. The Planetary Society. https://www.planetary.org/space-policy/nasas-fy-2025-budget. Accessed 19 June 0224.

PwC-UKSA. 2023. Expanding Frontiers: The Down to Earth Guide to Investing in Space. PwC with UK Space Agency. https://www.strategyand.pwc.com/uk/en/reports/expanding-frontiers-down-to-earth-guide-to-investing-in-space.pdf. Accessed 12 December 2023.

SIA. 2023. State of the Satellite Industry Report (SSIR). Satellite Industry Association (SIA). https://sia.org/record-setting-growth-highlights-commercial-satellite-industry-as-it-continues-to-dominate-expanding-global-space-business-sia-releases-26th-annual-state-of-the-satellite-industry-report/. Accessed 12 March 2024.

SIA. 2024. State of the Satellite Industry Report (SSIR). Satellite Industry Association (SIA). https://sia.org/commercial-satellite-industry-continues-historic-growth-dominating-global-space-business-27th-annual-state-of-the-satellite-industry-report/. Accessed 24 November 2024.

SIPRI. 2024. Global Military Spending Surges Amid War, Rising Tensions and Insecurity. Stockholm International Peace Research Institute. https://www.sipri.org/media/press-release/2024/global-military-spending-surges-amid-war-rising-tensions-and-insecurity . Accessed 12 May 2024.

Space Foundation. 2023. Space Foundation Releases The Space Report 2023 Q2, Showing Annual Growth of Global Space Economy to $546b. https://www.spacefoundation.org/2023/07/25/the-space-report-2023-q2/. Accessed 12 January 2024.

Space Foundation. 2024. The Space Report. https://www.thespacereport.org/resources/launch-records-topple-in-2024-with-busiest-january-of-space-age/. Accessed 12 July 2024.

Statista. 2023a. Toilet paper—Worldwide. https://www.statista.com/outlook/cmo/tissue-hygiene-paper/toilet-paper/worldwide.Accessed 20 January 2024.

Statista. 2024a. Leading Oil and Gas Companies Worldwide based on Revenue as of 2023. https://www.statista.com/statistics/272710/top-10-oil-and-gas-companies-worldwide-based-on-revenue/. Accessed 20 May 2024.

Statista. 2024b. Advertising Spending Worldwide. https://www.statista.com/outlook/amo/advertising/worldwide. Accessed 12 May 2024.

Signé, L. Dooley, H. 2023. How Space Exploration is Fueling the Fourth Industrial Revolution. Brookings. https://www.brookings.edu/articles/how-space-exploration-is-fueling-the-fourth-industrial-revolution/. Accessed 12 May 2024.

UBS. 2023. Global Wealth Report 2023: Leading Perspectives to Navigate the Future. https://www.ubs.com/global/en/family-office-uhnw/reports/global-wealth-report-2023.html. UBS. Accessed 14 March 2024.

USAF-FMC. 2024. UA Air Force Annual Budgets. US Department of the Air Force, Financial Management and Controller. https://www.saffm.hq.af.mil/FM-Resources/Budget/. Accessed 16 May 2024.

USSF. 2024. United States Space Force History. United States Space Force. https://www.spaceforce.mil/About-Us/About-Space-Force/History/#. Accessed 12 June 2024.

Washington Post. 2005. NASA's Griffin: 'Humans Will Colonize the Solar System'. https://www.washingtonpost.com/wp-dyn/content/article/2005/09/23/AR2005092301691.html. Washington Post. Accessed 12 May 2024.

WEF-McKinsey. 2024. Space: The $1.8 Trillion Opportunity for Global Economic Growth2024. World Economic Forum and McKinsey and Company. April 20204 Report. https://www3.weforum.org/docs/WEF_Space_2024.pdf. Accessed 10 May 2024.

3

Space and Outer Space

The scenery was very beautiful. But I did not see The Great Wall.
Yang Liwei, Shenzhou 5 Taikonaut, 2003

Peace be upon you.
Hazza Al Mansouri, Soyuz MS-15 Astronaut, 2019

Before discussing the challenges imposed by our financial value framework, mathematics, and monetary architecture, I believe it is critical to step back and address a very popular terminological conflation. Very often, the word space is used to refer to outer space and/or the terms are used interchangeably. In truth, this is why the title of the book refers to 'the race to space,' and the previous chapter started by introducing the 'space economy.' I wanted to make sure the subject matter is contextualised for a wider audience. This chapter is dedicated to fine-tuning our terminology, distinguishing between space and outer space, and offering a functional conceptualisation of space that can be used in the solutions introduced in later chapters. In other words, space and outer space are not synonymous, although outer space is a part of space.

The appropriate starting point of this discussion must be the self-description of the United Nations Committee on the Peaceful Uses of Outer Space (COPUOS). Using two different texts from the Committee's website, we can see how this terminological conflation is introduced. While its name refers to *outer space*, in its own text COPUOS refers to 'space.'

© The Author(s), under exclusive license to Springer Nature
Switzerland AG 2024
A. V. Papazian, *Financing the Race to Space*,
https://doi.org/10.1007/978-3-031-73102-0_3

The United Nations has been involved *in space activities* ever since the very beginning of *the space age*. Ever since the first human-made satellite orbited the Earth in 1957, the UN has been *committed to space* being used for peaceful purposes. (COPUOS 2024a)
The Committee on the Peaceful Uses of Outer Space (COPUOS) was set up by the General Assembly in 1959 to govern the exploration and *use of space* for the benefit of all humanity: for peace, security and development. The Committee was tasked with reviewing international cooperation in *peaceful uses of outer space*, studying *space-related activities* that could be undertaken by the United Nations, encouraging *space research programmes*, and studying legal problems arising from the *exploration of outer space*. (COPUOS 2024b)[1]

Meanwhile, the Outer Space Treaty (OST) is more careful in its use of the terms. The preamble and first article clearly refer to outer space. Outer Space Treaty is a shorter name for the Treaty on Principles Governing the Activities of States in the Exploration and Use of Outer Space, including the Moon and Other Celestial Bodies (Outer Space Treaty), adopted by the UN General Assembly resolution 2222 (XXI) on the 19th of December 1966, which entered into force on the 10th of October 1967.

Inspired by the great prospects opening up before mankind as a result of man's entry into outer space.... *Article I:* The exploration and use of outer space, including the Moon and other celestial bodies, shall be carried out for the benefit and in the interests of all countries, irrespective of their degree of economic or scientific development, and shall be the province of all mankind. Outer space, including the Moon and other celestial bodies, shall be free for exploration and use by all States without discrimination of any kind, on a basis of equality and in accordance with international law, and there shall be free access to all areas of celestial bodies. There shall be freedom of scientific investigation in outer space, including the Moon and other celestial bodies, and States shall facilitate and encourage international cooperation in such investigation. (UNOOSA 1966)

The treaty refers to outer space including the moon and other celestial bodies, not 'space.' Meanwhile, looking at definitions of space, the Merriam-Webster dictionary, amongst other definitions, describes space as "a boundless three-dimensional extent in which objects and events occur and have relative position and direction; a physical space independent of what occupies it" (Merriam-Webster 2024).

---

[1] Emphasis added to denote how 'space' and 'outer space' are used interchangeably, despite the very name of the Committee.

The Oxford English Dictionary provides a number of definitions including: "Linear distance; Interval between two or more points, objects, etc.; Physical extent or area; Extent in two or three dimensions; The expanse in which celestial objects are situated; The physical universe (excluding celestial objects) beyond the earth's atmosphere, consisting of near vacuum with small amounts of gas and dust; Continuous, unbounded, or unlimited extent in every direction, without reference to any matter that may be present; The physical expanse which surrounds something; Extent in all directions from a given point or object" (EOD 2024).

For its 1.5 billion unique visitors per month Wikipedia defines outer space as follows—the text in parenthesis reveals this conflation:

Outer space (or simply space) is the expanse beyond celestial bodies and their atmospheres. It contains ultra-low levels of particle densities, constituting a near-perfect vacuum of predominantly hydrogen and helium plasma, permeated by electromagnetic radiation, cosmic rays, neutrinos, magnetic fields and dust. (Wikipedia 2024)

The point of this introduction is to establish the necessity to clarify what we mean by space. Indeed, assuming that 'space economy' means 'outer space economy' is not very helpful or very accurate. In physical terms, outer space is part of space, as space does not end at the edge of the planet's atmosphere, and our planet is in space. In other words, we need to start by establishing terminological clarity. In the following sections I offer a definition and conceptualisation of space as our all-encompassing physical context of matter.

3.1 Space and Outer Space

Matter and energy are the two basic components of the entire Universe. An enormous challenge for scientists is that most of the matter in the Universe is invisible and the source of most of the energy is not understood. (Center for Astrophysics 2024)

In the above quote invisible matter is dark matter, and the energy that is not yet understood is dark energy. The Center for Astrophysics | Harvard & Smithsonian suggests that "95% of matter and energy in the Universe is currently unobservable" (Center for Astrophysics 2024). I am starting with this description of the content of the Universe because it is important to establish that our understanding is still incomplete. While we have made immense progress in understanding and theorising the nature of our reality and the forces that shape it, this understanding has many gaps and many paradoxes.

Naturally, this book is not about physics and the many debates within the field (Smolin 2006; Greene 2004; Rovelli 2018). While a curious enthusiast, I cannot and do not approach this topic from a theoretical physics perspective. However, the definition of space I provide in this book, and have discussed in previous publications (Papazian 2023, 2022), is informed and inspired by the works that have shaped our understanding of our physical context.

In physics and philosophy, the concepts of matter, energy, space, time, and spacetime have been the subject of extensive debate. Nicolaus Copernicus, Johannes Kepler, Galileo Galilei, Isaac Newton, Christiaan Huygens, Gottfried Wilhelm Leibniz, Hermann Minkowski, Albert Einstein, Niels Bohr, Werner Heisenberg, Paul Dirac, Max Planck, and Erwin Schrödinger are some of the giants who have shaped our understanding of space, time, energy, and matter on the level of subatomic particles as well as planets. For some, consciousness comes into play. Max Planck, a theoretical physicist who won the Nobel Prize in Physics (1918) states the following in an interview with The Observer in 1931:

> I regard consciousness as fundamental. I regard matter as derivative from consciousness. We cannot get behind consciousness. Everything that we talk about, everything that we regard as existing, postulates consciousness. (Observer 1931)

Naturally, and however interesting, philosophies of consciousness are also outside the scope of this book. The key point here is that we have been and are still in the process of thinking and interpreting the universe, trying to make sense of matter and the forces that govern our physical reality. Exploring what, where, and how we are where and what we are is the mark of a living civilisation.

Given that this book is about our financial and monetary reality, I focus the discussion on the tangible physical context that we find ourselves in, that we interpret and express ourselves through, that we shape with our daily productive and non-productive activities.

I define space as our physical context of matter, irrespective of constitution, composition, density, dynamics, and temperature, stretching from subatomic to interstellar space and every layer in between and beyond, where outer space is but a segment. In Fig. 3.1 I offer a layered conceptualisation of space, depicting layers of space we have come in direct contact with physically or through our rovers and probes and other technologies (Papazian 2022).

I define space as our physical context of matter *irrespective of constitution, composition, density, dynamics,* and *temperature* because different layers have very different characteristics when it comes to these aspects. This reflects the observation we started with regarding dark matter. Similarly, this reflects the fact that besides having different constitution, composition, and density, these layers can also be governed by different dynamics, like quantum or classical. Moreover, they can have similar composition but different temperature and different density.

Unlike other versions of this layered conceptualisation of space (Papazian 2023, 2022), Fig. 3.1 also introduces the human body, the datasphere, cislunar space, and Mars as additional layers. The human body is recognised as a unique layer in itself with different characteristics and exposure. Given that the human body in zero gravity and the human body on the continental

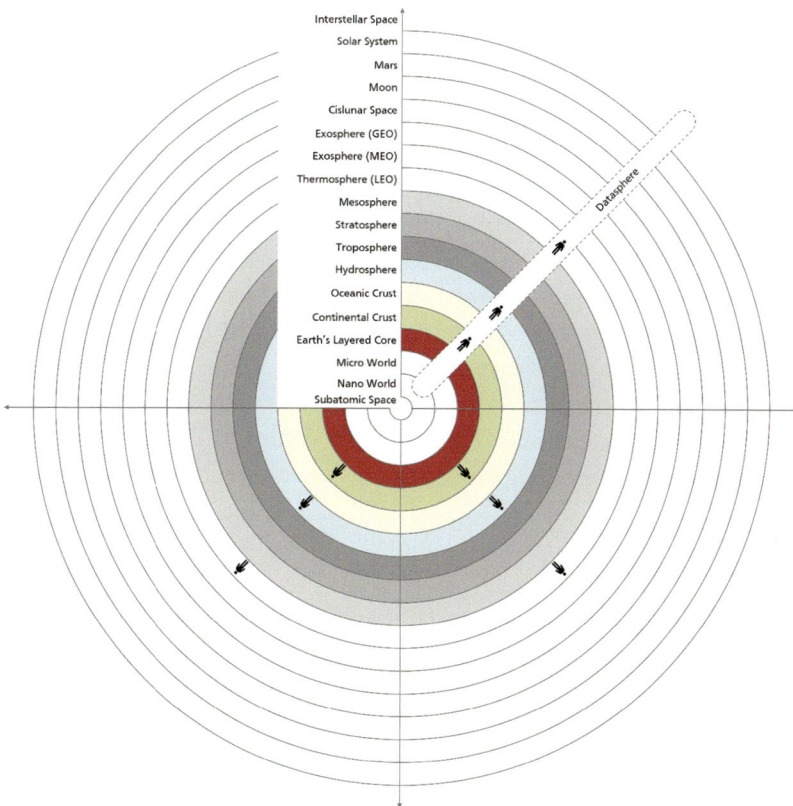

Interstellar Space
Solar System
Mars
Moon
Cislunar Space
Exosphere (GEO)
Exosphere (MEO)
Thermosphere (LEO)
Mesosphere
Stratosphere
Troposphere
Hydrosphere
Oceanic Crust
Continental Crust
Earth's Layered Core
Micro World
Nano World
Subatomic Space

Datasphere

Fig. 3.1   Space layers (*Source* Adapted and Updated from Papazian [2022])

crust, or on/under water, cannot be treated as the same, the location of the human body is relevant.

Figure 3.1 identifies the continental crust, the hydrosphere, and the thermosphere as the three space layers where humans have permanent physical presence. The continental crust is where nearly the entire species lives, the hydrosphere is where human built and navigated ships are a constant presence, and the thermosphere is where humanity's two space stations are, the International Space Station (ISS) and the Tiangong Space Station (TSS). Although we access the stratosphere every day through domestic and international travel, our airplanes do not stay there for continuous and long stretches of time. In other words, we transit the stratosphere, but we are not a permanent presence, even if we most definitely have airplanes flying in the layer at any given moment. Indeed, the longest direct flight covers a distance of around 9,537 miles lasting an estimated 18 hours and 50 minutes.

The datasphere is where and through which we interact, exchange memories, emotions, information, goods, and services in the form of digital data. It stretches from the nano world to interstellar space and beyond. This is so because the microchips that enable their storage (electronic circuits on silicon) are extremely small and are measured in nanometre (one billionth of a meter), and the data can be received from probes in interstellar space like Voyager 1, which reached interstellar space on the 25th of August 2012 (JPL-NASA 2024). Moreover, given submarine fibre optic cables that make much of our connectivity possible, datasphere also goes through the hydrosphere.

Figure 3.1 also includes Mars as an added layer given the commonly held view that we should go to Mars. It also introduces cislunar space as an additional layer relevant to the growing activities in outer space. A recent report by the Office of Science and Technology Policy (OSTP) within the Executive Office of the President of the United States is entirely dedicated to National Cislunar Science & Technology Strategy. This strategy paper is preceded by the 'Primer on Cislunar Space' written by Holzinger, Chow, and Garretson (2021), and the NASA and US Space Force MOU where reference to Cislunar space is prominent (NASA-USSF 2020).

> Cislunar space, the three-dimensional region of space beyond Earth's geosynchronous orbit but still within the gravitational influence of the Earth and/or the Moon, is a new sphere of human activity with diverse endeavors accelerating in the decade ahead. Cislunar space offers tremendous promise for advancing science, technology, and exploration. (OSTP 2022, 2)

In a parallel strategy report published in March 2023, the Office of Science and Technology Policy (OSTP 2023) revealed the importance of a national strategy for Low Earth Orbit (LEO) research and development. Naturally, on our way to the moon and Mars, and as the most active layer of outer space, Low Earth Orbit is strategically important.

> Innovative public-private collaborations and increased commercial access to LEO have sparked a growing interest in space research and development. The increased potential for rapid discoveries is expanding the commercial market at the same time the International Space Station (ISS) is beginning its phased retirement to be replaced by lower cost commercial alternatives. As a result of this changing landscape, the U.S. Government needs to plan strategically for the post-ISS world especially for the employment of commercial state-of-the-art autonomously-operated and human-occupied outposts. (OSTP 2023, 5)

In Fig. 3.1 the micro world, nano world, and subatomic space layers are intentionally at the centre of the other layers. This is because they are all available and accessible from any point of matter. Naturally, one cannot directly explore the molecules and atoms inside the Earth's layered core, which includes the upper mantle, the lower mantle, outer core, and inner core, but the micro fabric of matter is present across all the identified layers, including the human body.

The space layers depicted in Fig. 3.1 can, of course, be broken down into sublayers, and sublayers can be defined into further layers. Indeed, outer space is a vast landscape far beyond interstellar space, including our galactic context and the many billions of galaxies in the observable universe. The purpose of the space layers diagram is to be a high-level representation of our physical context, depicting the layers of space we have come into direct or indirect contact with.

This conceptualisation of space is necessary and a foundational element of the solutions I offer in later chapters. To recap, space is our all-encompassing physical context of matter, irrespective of constitution, composition, density, dynamics, and temperature, stretching from subatomic to interstellar space and every layer in between and beyond, where outer space is but a segment made up of many layers.

3.2   Conceptual Lines in Space

The layered conceptualisation of our physical context, of space, is a *conceptualisation*. Just like all our conceptual inventions that help us structure and navigate our physical context of matter, space layers can help us structure our context. Indeed, as mentioned, the conceptualisation presented in Fig. 3.1 plays a central role in the analytical solutions I offer in later chapters. To expand and explain this relevance, I discuss other conceptual (imaginary) lines that help us understand, structure, and organise ourselves and our activities on planet Earth, in space.

3.2.1   Country Borders

The first most obvious example of conceptual lines that help us define ourselves are country borders. Anyone who has travelled from one country to another knows, whatever the means of transportation, these borders are human projections. Their demarcation and physical manifestation are not natural or geographic, but political and man-made. This is true even for island countries defined and self-contained by their shores. These lines/borders are conceptually projected on the surface of the planet by us. Whatever their historic origin, longevity, and heritage, and the level of pride they may or may not inspire, these borders do not cease to be imaginary conceptual lines projected by us.

Heavily defended and entrenched in local culture and institutional structures, country borders and their history are human interpretations and projections. Figure 3.2 depicts the world political map, which defines human existence on the surface of our planet. These borders shape and influence humanity's political, sociocultural, and economic existence. Indeed, many humans interpret themselves and their existence in relation to this map, building their identity upon imaginary lines in space, and the sociocultural history that surrounds their development over time.

The point here is to recognise the role that our imaginary lines play, and how deep and defining their impact can be. This is true for individuals, collectives, and humanity as a whole. Figure 3.2 contextualises the political map with the space layers diagram in order to draw attention to the conceptual and physical location of the map within the wider and much grander context of space.

The impact of this map is not just on the surface. These imaginary lines dictate differences and limitations the human body, on the continental crust, the oceanic crust, the hydrosphere, and the atmosphere. As the next section will reveal, where national airspaces end and outer space begins is defined by yet another imaginary line.

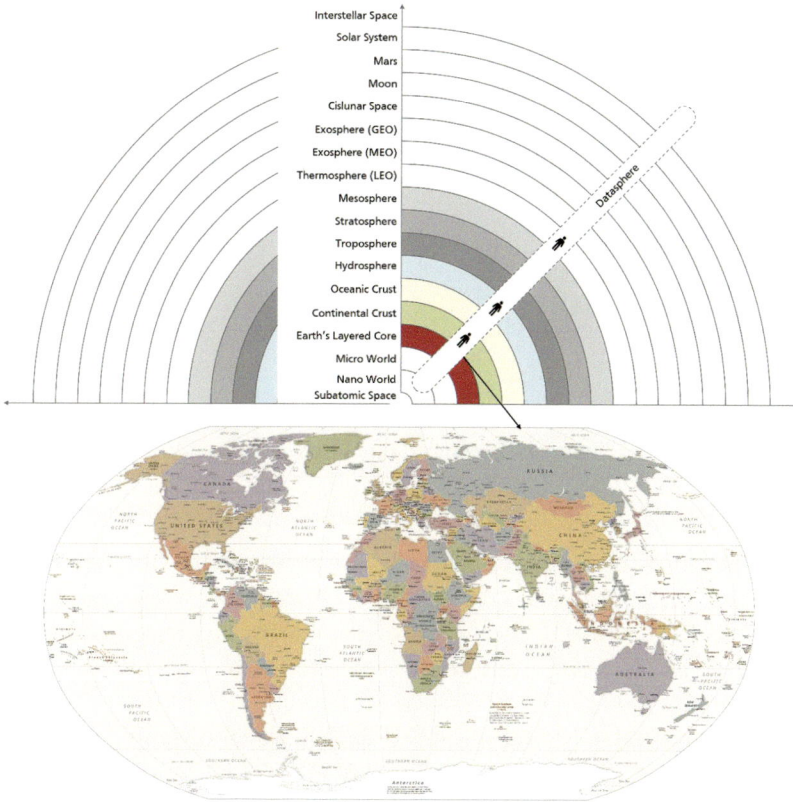

Fig. 3.2 World political map and space layers (*Source* Author using World Map from CIA [2023])

3.2.2 The Kármán Line

Another important conceptual line is the Kármán line. It defines the end of national airspace and the beginning of outer space. This is an important imaginary line. It does not exist in physical form, and yet it helps identify the edge of national airspaces across the map. The Kármán line, 62 miles or 100 km above sea level, was coined by Andrew G. Haley and named after Theodore von Kármán, inspired through his work on defining the theoretical boundary between aeronautics and astronautics. The main purpose and implications of the Kármán line are legal and regulatory.

Figure 3.3 (not to scale) depicts the Kármán line and shows something quite interesting. We know that the *average* depth of the continental crust is around 30 km, and the Kármán line is 100 km above sea level. This means

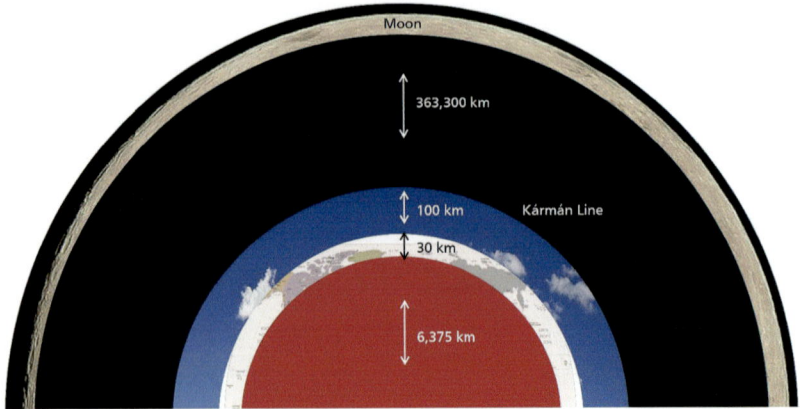

Fig. 3.3 The Kármán line (not to scale) (*Source* Author, using World Map from CIA [2023])

that the world as we know it, defined by the imaginary lines of the political map on the continental crust and the Kármán line above, is on average 130 km thin. Meanwhile, the radius of the Earth, digging into the Earth's layered core, is 6,375 Km,[2] and the distance to the moon at its closest point to Earth is 363,300 km.

In other words, despite its interpretive dominance, our political map-based reality applies to a tiny fraction of the physical continuum of space that we live in, a 130 km (80.8 miles) bandwidth in an astronomically large context. Interestingly, the distance between Houston and Dallas is 363 km.

The conceptual line that defines the starting point of outer space (the Kármán Line) and the imaginary lines that define countries (borders on map) on the continental crust, and their corresponding national airspaces, shape much of our civilisational content today. From economic, monetary, and financial perspectives, these lines distinguish and demarcate our markets and investments, and the landscape of rules and regulations that must be followed.

The futurist in me would like to open a parenthesis and argue that, over time, our growing awareness of space will naturally shift our understanding and interpretation of ourselves to the vertical axis of our existence, which is astronomically immense. Eventually, defining ourselves through the political map, which has an interpretive reach of 130 kilometres on average, will become an unsatisfactory equation, leading to changes in the lines and structures that shape our lives on the surface.

---

[2] Earth's layered core includes the Upper Mantle, the Lower Mantle, Outer Core, and Inner Core.

3.2.3   The Prime Meridian and the Equator

Yet another conceptual and imaginary line that shapes our understanding of terrestrial space is the Prime Meridian. Epitomised by the Greenwich laser beam, at 0° longitude, the Prime Meridian has been and is a fundamental pillar of human civilisation. It helps us structure our days and activities and allows us to navigate our terrestrial environment (See Fig. 3.4).

> The Prime Meridian is the line and the point at which the world's longitude is set at 0°. It does not exist in any strict material sense, yet through maps and clocks, the prime meridian governs the life of every human on Earth. (Withers 2017, 5)

This imaginary point/line on Earth, in space, which defines our maps and our clocks, acts as a structural pillar of the entire world economy. Our clocks and navigation tools would stop working without the prime meridian. And yet, it is conceptually projected onto our physical context, it does not actually exist except through human convention. The same is true for the equator which divides the Earth into northern and southern hemispheres at 0° latitude. I will revisit the Prime Meridian and its history in chapter eight, for now, a quick note on Null Island is due.

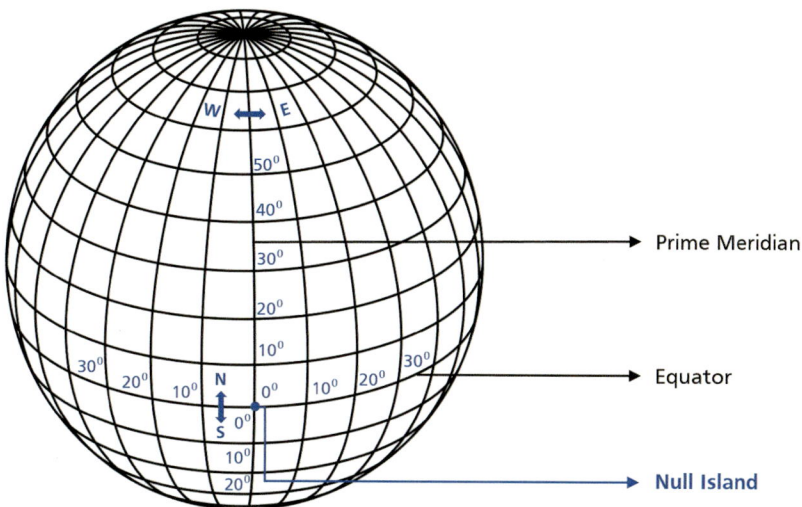

Fig. 3.4   Prime meridian, equator and Null Island (*Source* Author)

3.2.3.1 Null Island

Linked to the Prime Merdian and the Equator is the concept of the Null Island, the location at zero degrees latitude and zero degrees longitude (0°N 0°E), i.e., where the prime meridian and the equator intersect. An imaginary island in the middle of the Gulf of Guinea in the southeast Atlantic Ocean that does not exist, and yet acts as a central pillar of a framework that governs our geographic information systems (GIS), and therefore our entire productive life through maps and clocks (St Onge 2016).

3.2.4 Electron 'Orbits'

Yet more imaginary conceptual lines can be found in our understanding of matter on the nanoscale. Atoms, which typically have diameters between 0.1 and 0.5 nanometre, are often described and drawn as perfectly rational and precise structures. Figure 3.5 depicts a standard drawing of the Carbon atom with 6 protons and 6 neutrons in the atomic nucleus and 6 orbiting electrons.

Interestingly, electron orbits or orbitals, as we have come to learn through quantum physics, are not really lines, but probability clouds. A historic contribution that defined the trajectory of our understanding of matter and the atom, and the behaviour of electrons, is Heisenberg's famous uncertainty

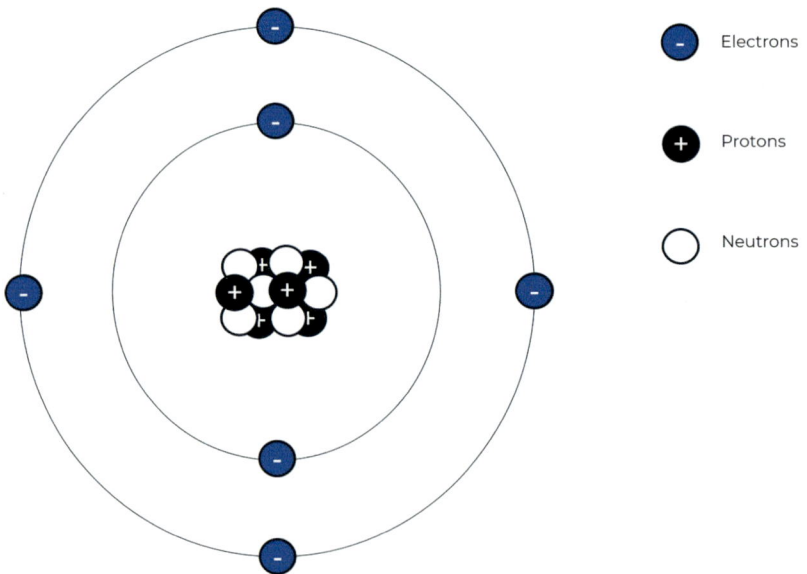

Fig. 3.5 Carbon atom structure (*Source* Author)

principle. Below is a direct quote from Heisenberg's 1927 article translation (1983) made available by NASA.

> Let us move on to the concept of the 'path of the electron.' By path or trajectory we mean a series of points in space (in a given reference system) that the electron adopts as successive 'positions.' Since we already know what 'position at a certain time' means, there are no new difficulties, here. It is still readily understood that the often used expression, for instance, 'the 1-S orbit of the electron in the hydrogen atom' makes no sense, from our point of view. Because in order to measure this 1S orbit, we would have to illuminate the atom with light such that its wavelength is considerably shorter than 10^{-8} cm. But one light quantum of this kind of light would be sufficient to completely throw the electron out of its 'orbit' (for which reason never more than a single point of this 'path' could be defined, in space) and hence the word 'path' is not very sensible or meaningful, here. This can be easily derived from the experimental possibilities, even without any knowledge of the new theories. (Heisenberg 1927, 6)

We have now identified different kinds of human conceptual lines that define and structure our physical context of matter, our space. These lines do not exist in real terms, and yet, they are all pillars of our productive and non-productive activities on Earth, in space. What would happen to human civilisation if we were to remove these lines? Would there be more chaos than we already have? Can we improve or add on our imaginary lines to serve us and our evolution better? I believe we can, and we should.

3.3   Mapping Lunar and Mars Habitats

Now that we have established that much of our mapping and understanding of space is built upon conceptual or imaginary lines, let me revisit the conceptualisation of space we discussed at the beginning of this chapter. Figure 3.6 is identical to Fig. 3.1 with the additional notations identifying a permanent human presence on the moon and Mars.

This conceptualisation of our physical context provides the framework upon which the solutions I offer are based. Indeed, if the propositions introduced in later chapters are adopted, this would be an appropriate representation of where we would be after implementation.

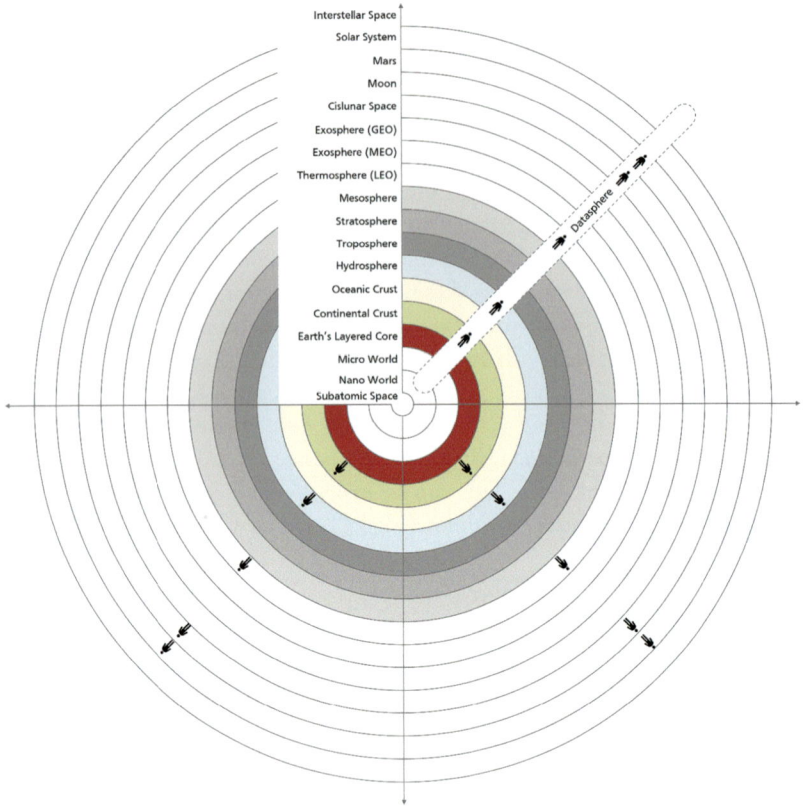

Fig. 3.6 Space layers with lunar and mars habitats (*Source* Adapted and Updated from Papazian [2022])

3.4 Conclusion

This chapter was dedicated to fine-tuning our terminology and offering a new conceptualisation of space, our physical context of matter. Given that much of the recent literature on the outer space economy refers to 'the space economy' or to the 'new space race,' this distinction was necessary. From this point onward, outer space is a segment of space, and the words are not interchangeable.

I offered a layered conceptualisation of space, our physical context of matter, irrespective of constitution, composition, density, dynamics, and temperature, stretching from subatomic space to interstellar space and every layer in between and beyond, where outer space however vast, is but a segment.

This layered conceptualisation of our physical context of matter is a foundational element of the framework and solutions I offer in later chapters. Moreover, it is a conceptualisation, just like the many other imaginary or conceptual lines that help us make sense of our physical context, like country borders, the Kármán line, the prime meridian, the equator, and electron orbits. Furthermore, the start and end points of the space layers are also human inventions, and the layers can be broken down into sublayers, and sublayers can be further defined when and if necessary.

While I will come back to the Prime Meridian and the layered conceptualisation of space later on in the discussion, in the next chapter I discuss humanity's treatment of space, including outer space. As we will see, and can observe around us today, our impact on space reveals a dismal picture.

References

Center for Astrophysics. 2024. What is the Universe Made Of? Centre for Astrophysics, Harvard & Smithsonian. https://www.cfa.harvard.edu/big-questions/what-universe-made. Accessed 12 June 2024.

CIA. 2023. World Policial Map—The World Factbook. Washington, DC: Central Intelligence Agency, 2023. https://www.cia.gov/the-world-factbook/static/7236711e3e48e1a71e1210107f15e7b4/world_pol.pdf. Accessed on 12 May 2024.

COPUOS. 2024a. COPUOS History. COPUOS. https://www.unoosa.org/oosa/en/ourwork/copuos/history.html. Accessed 24 June 2024.

COPUOS. 2024b. Committee on the Peaceful Uses of Outer Space. COPUOS. https://www.unoosa.org/oosa/en/ourwork/copuos/index.html. Accessed 24 June 2024.

EOD. 2024. Space. Oxford English Dictionary. https://www.oed.com/dictionary/space_n1. Accessed 22 June 2024.

Greene, B. 2004. *The Fabric of the Cosmos: Space, Time, and the Texture of Reality*. London: Penguin Books.

Heisenberg, W. 1927. The Actual Content of Quantum Theoretical Kinematics and Mechanics. Translation, from NASA Archive. December 1983. https://ntrs.nasa.gov/api/citations/19840008978/downloads/19840008978.pdf. Accessed 12 March 2024.

Holzinger1, M. J., Chow, C.C., Garretson, P. 2021. A Primer on Cislunar Space. US Air Force Research Laboratory. https://www.afrl.af.mil/Portals/90/Documents/RV/A%20Primer%20on%20Cislunar%20Space_Dist%20A_PA2021-1271.pdf. Accessed 12 May 2024.

JPL-NASA. 2024. Interstellar Mission. NASA. https://voyager.jpl.nasa.gov/mission/interstellar-mission/. Accessed 16 April 2024.

Merriam-Webster. 2024. Space. https://www.merriam-webster.com/dictionary/space. Accessed 10 April 2024.

NASA. 2024. NASA's Voyager 1 Resumes Sending Engineering Updates to Earth. Jet Propulsion Laboratory. NASA. https://www.jpl.nasa.gov/news/nasas-voyager-1-resumes-sending-engineering-updates-to-earth. Accessed 12 May 2024.

NASA-USSF. 2020. Memorandum of Understanding between the National Aeronautics and Space Administration and the United States Space Force. NASA. https://www.nasa.gov/wp-content/uploads/2015/01/nasa_ussf_mou_21_sep_20.pdf?emrc=a97009. Accessed 12 March 2024.

Observer. 1931. Interviews with Great Scientists. The Observer. https://www.newspapers.com/article/the-observer-the-observer-sunday-janua/121386587/. Accessed 12 March 2024.

OSTP. 2023. National Low Earth Orbit Research and Development Strategy. Low Earth Orbit Science and Technology Interagency Working Group and National Science and Technology Council. Office of Science and Technology Policy. https://www.whitehouse.gov/wp-content/uploads/2023/03/NATIONAL-LEO-RD-STRATEGY-033123.pdf. Accessed 12 March 2024.

OSTP. 2022. National Cislunar Science & Technology Strategy. Cislunar Technology Interagency Working Group and National Science and Technology Council. Office of Science and Technology Policy. https://www.whitehouse.gov/wp-content/uploads/2022/11/11-2022-NSTC-National-Cislunar-ST-Strategy.pdf. Accessed 12 March 2024.

Papazian, A. 2022. *The Space Value of Money: Rethinking Finance Beyond Risk and Time*. New York: Palgrave Macmillan. https://doi.org/10.1057/978-1-137-594 89-1.

Papazian, A. 2023. *Hardwiring Sustainability into Financial Mathematics: Implications for Money Mechanics*. New York: Palgrave Macmillan. https://doi.org/10.1007/978-3-031-45689-3.

Rovelli, C. 2018. *The Order of Time*. New York: Riverhead Books.

Smolin, L. 2006. *The Trouble with Physics*. Penguin Books.

St Onge, T. 2016. The Geographical Oddity of Null Island. Library of Congress. https://blogs.loc.gov/maps/2016/04/the-geographical-oddity-of-null-island/. Accessed 12 March 2024.

UNOOSA. 1966. Treaty on Principles Governing the Activities of States in the Exploration and Use of Outer Space, including the Moon and Other Celestial Bodies. United Nations Office for Outer Space Affairs. https://www.unoosa.org/pdf/gares/ARES_21_2222E.pdf. Accessed 10 May 2024.

Wikipedia. 2024. Outer Space. https://en.wikipedia.org/wiki/Outer_space. Accessed 12 April 2024.

Withers, W.J.C. 2017. *Zero Degrees: Geographies of the Prime Meridian*. Cambridge: Harvard University Press.

4

Human Impact on Space

This planet is not terra firma. It is a delicate flower and it must be cared for. It's lonely. It's small. It's isolated, and there is no resupply. And we are mistreating it.
Scott Carpenter, Mercury Aurora 7 Astronaut, 1962

In this immensity teeming with billions and billions of planets similar to ours, our disappearance would go unnoticed.
Jean-Loup Chrétien, Soyuz T-6 Astronaut, 1982

Having refined our terminology and presented a layered conceptualisation of space, in this chapter I provide a snapshot of our impact in and on space. Naturally, it is impossible to cover all aspects of human impact in one chapter. Furthermore, given that this is not the main theme of the book, the discussion is brief. The purpose is to showcase a number of key features contextualised by the space layers defined in the previous discussion.

Space is where all human productive and non-productive activities take place. It includes our bodies, our oceans, our lands, our atmosphere, and outer space and acts as the context through and within which we define humanity and human civilisation. In other words, we must realise and accept that the disappointing picture described in the following sections is of our own making. Ultimately, as Carl Sagan put it so eloquently, we are the ones who must save us from ourselves.

A. V. Papazian, *Financing the Race to Space*, https://doi.org/10.1007/978-3-031-73102-0_4

Our posturings, our imagined self-importance, the delusion that we have some privileged position in the Universe, are challenged by this point of pale light. Our planet is a lonely speck in the great enveloping cosmic dark. In our obscurity, in all this vastness, there is no hint that help will come from elsewhere to save us from ourselves. (Sagan 1994)

To do so we must address the widely documented ecological and environmental crises we face today. To do so, we must rethink our impact in space. Indeed, our inability to invest and build a sustainable and fair reality on Earth, just like our inability to invest and expand our reach in outer space, is linked to and caused by our current financial value framework, financial mathematics, and monetary architecture.

While I elaborate this argument in later chapters, in the following sections I briefly present human impact on space and its many layers.

4.1   Impact on Space Layers

Human productive and non-productive activities affect many different layers of space (Fig. 4.1). While we operate mainly from the continental crust, our reach goes far beyond. Our impact on space, our physical context of matter, irrespective of constitution, composition, density, dynamics, and temperature, stretching from subatomic space to interstellar space and every layer in between and beyond, is astonishingly careless and irresponsible. I discuss a select number of well-documented aspects to demonstrate that we have littered every environment we have come to touch. Human productivity is evidently unconcerned with our evolutionary continuity and quality of life.

4.1.1 Atmosphere: Climate Change

The most obvious and most frequently discussed aspect of human impact is on our atmosphere. Climate change and global warming due to Green House Gas emissions (GHG) have been front-page news due to the many floods, extreme hailstorms, wildfires, and droughts. While some are affected more severely than others, the symptoms are global, and the examples are too many to list. The evidence confirming human responsibility for climate change has been overwhelming and well documented (IPCC 2023, 2022, 2021, 2018, 2013, 2007). IPCC (2023) summarises the challenge as follows:

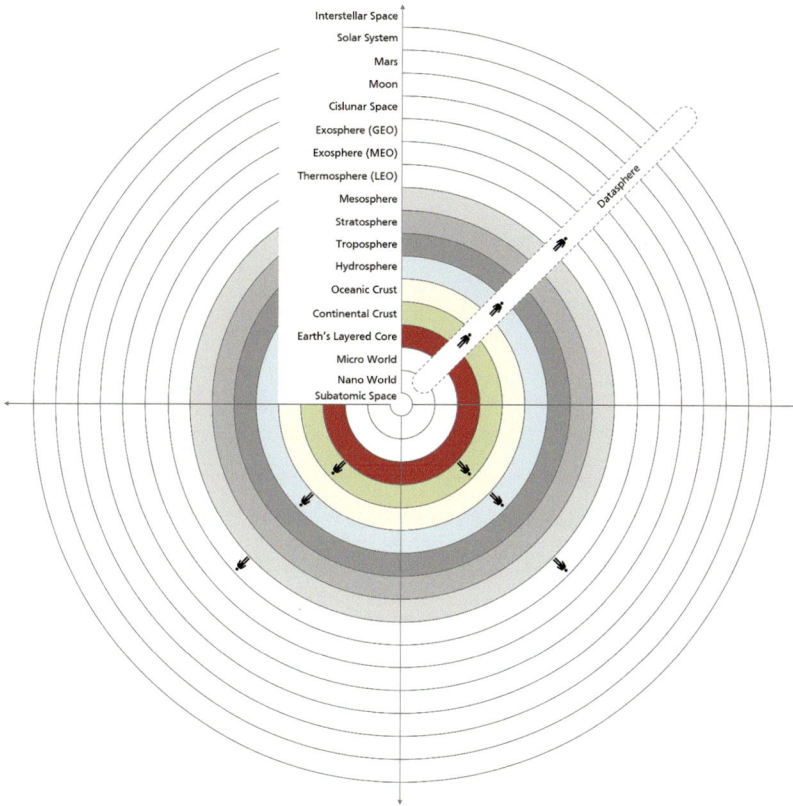

Fig. 4.1 Space layers (*Source* Adapted and Updated from Papazian [2022])

Human activities, principally through emissions of greenhouse gases, have unequivocally caused global warming, with global surface temperature reaching 1.1°C above 1850–1900 in 2011–2020. Global greenhouse gas emissions have continued to increase, with unequal historical and ongoing contributions arising from unsustainable energy use, land use and land-use change, lifestyles and patterns of consumption and production across regions, between and within countries, and among individuals. (IPCC 2023, 4)

The Paris Agreement (UNFCCC 2015), which came into force in 2016, established a legally binding commitment to reduce GHG emissions. The target, 'to keep world temperature increases below 2 °C above pre-industrial levels and ideally limit the temperature increase to 1.5 °C,' has become a defining theme across many fields and industries. It has led to a diverse set of efforts, frameworks, and standards that aim to entrench this objective into the business and financial operations of firms and governments.

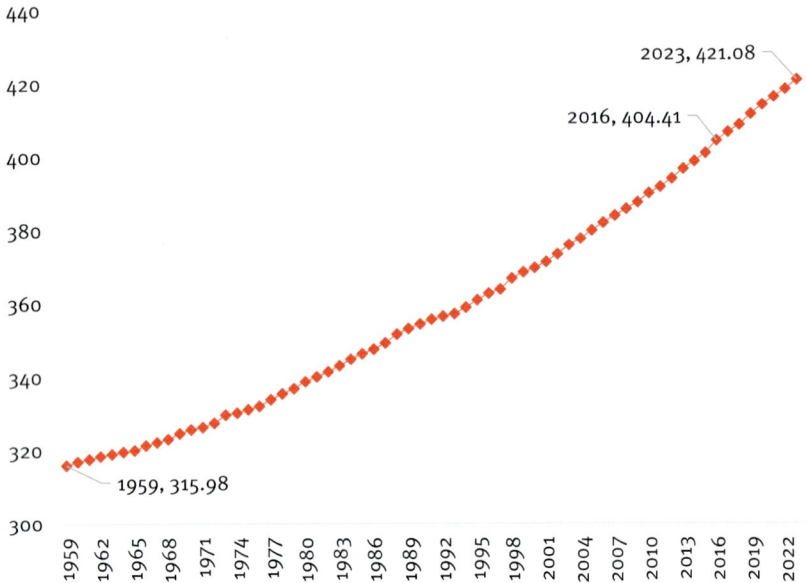

Chart 4.1 Atmospheric CO$_2$ levels, yearly mean, mole fraction in dry air (ppm) (*Source* NOAA [2024])

In spite of all the legal agreements, the frameworks, the standards, the reporting requirements, and the flood of media opinions, we have not changed our trajectory yet. As depicted in Chart 4.1, atmospheric CO$_2$ levels are still going up (NOAA 2024). Ironically, despite the ambition and the rhetoric, fossil fuel subsidies are also rising and were at a record $7 trillion in 2022 (IMF 2023) (see Chart 4.2). In the same spirit, the UK government recently approved hundreds of new licenses for oil and gas exploration in the North Sea (UK Government 2023). In July 2024, Copernicus released its monthly climate bulletin where it revealed that global temperatures have already reached and crossed the 1.5 °C above pre-industrial level since July 2023 (Copernicus 2024) (see Chart 4.3). To put it simply, we continue to destroy our atmosphere despite the evidence.

4.1.2   Hydrosphere: Plastic in Our Oceans

There are an estimated 50–75 trillion pieces of plastic and microplastics in our oceans (IOC-UNESCO 2022). Pew Charitable Trusts (PEW 2020) estimates that 11 million metric tons of plastic waste entered the oceans in 2016, and the number is set to rise to 29 million metric tons per year by 2040. PEW (2020) provides a chilling summary of the untenable trajectory as follows:

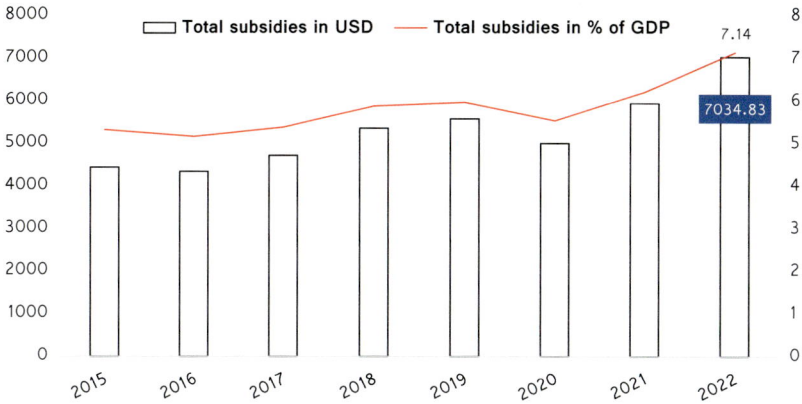

Chart 4.2 Fossil fuel subsidies in $Billion (left axis) and % (right axis) (*Source* IMF [2023])

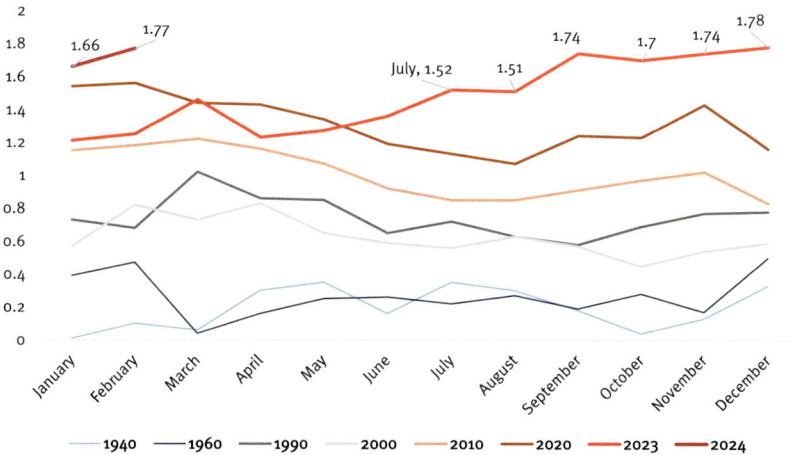

Chart 4.3 Monthly global surface air temperature anomaly relative to 1850–1900, in C° (*Source* Copernicus [2024])

We estimate that 11 million metric tons of plastic entered the ocean from land in 2016, adding to the estimated 150 million metric tons of plastic already in the ocean. Plastic flows into the ocean are projected to nearly triple by 2040 to 29 million metric tons per year. Even worse, because plastic remains in the ocean for hundreds of years, or longer, and may never biodegrade, the cumulative amount of plastic stock in the ocean could grow by 450 million metric tons in the next 20 years— with severe impacts on biodiversity, and ocean and human health. (PEW 2020, 25)

As described in Chart 4.4, the above is the direct result of an ever-increasing amount of plastic waste generation. PEW (2020) estimates that in 2030 and 2040, under business-as-usual conditions, we would be generating 330.4 and 419.7 million metric tons, respectively.

These figures have had a real impact on our oceans. We now have 5 offshore plastic accumulation zones. The largest amongst them is the Great Pacific Garbage Patch (GPGP), which is somewhere between Hawaii and California (Ocean Cleanup 2024). Figure 4.2 identifies the approximate location of the five zones (the figure is not to scale).

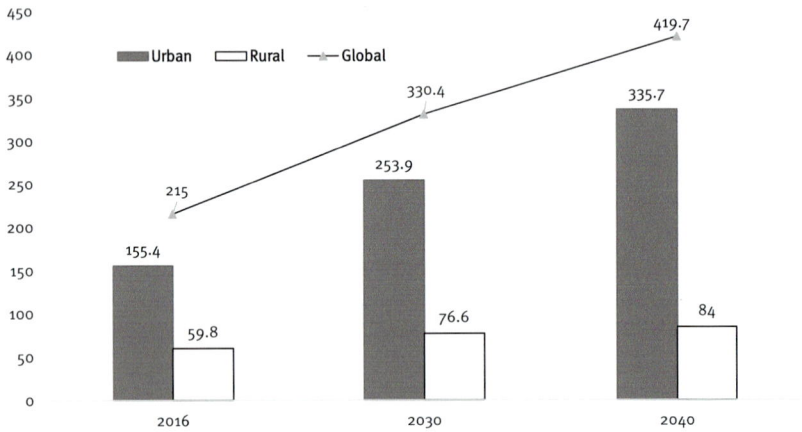

Chart 4.4 Total annual plastic waste generated in million metric tons (projections under business-as-usual conditions) (*Source* PEW [2020])

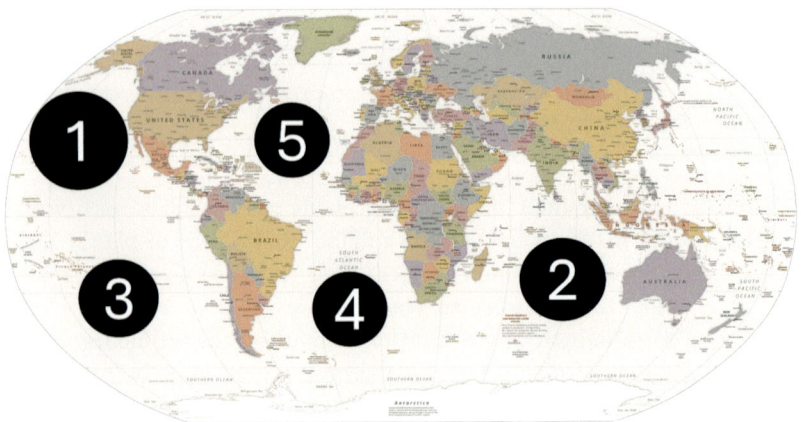

Fig. 4.2 The five offshore plastic accumulation zones in the world's oceans (*Source* Author, using Ocean Cleanup [2024], and map from CIA [2023])

I have intentionally used the political map to depict the five zones. The political map, as discussed in Chapter 3, plays a prominent role in humanity's self-interpretation at the moment, but it does not show these islands of garbage and waste that this very civilisation has been creating. The Great Pacific Garbage Patch covers an estimated surface area of 1.6 million square km, "twice the size of Texas or three times the size of France" (Lebreton et al. 2018; Ocean Cleanup 2024). As such, it is my view that all future political maps should show these garbage patches as an authentic civilisational representation.

4.1.3   Micro World and Human Body: Pollutants in Our Lungs

Based on estimates by the European Environment Agency (EEA 2023) fine particulate matter ($PM_{2.5}$) have significant impacts on human health and are one of the major causes of premature death in Europe. $PM_{2.5}$ and/or PM_{10} are airborne particulate matter (PM) that represent a group of pollutants, including different kinds of chemical substances. In a recent UK Government (2024) publication this focus on measuring PM is explained and justified due to their health implications.

> Particulate matter (PM) is everything in the air that is not a gas. It consists of a huge variety of chemical compounds and materials, some of which can be toxic. Due to the small size of many of the particles that form PM some of these toxins may enter the bloodstream and be transported around the body, lodging in the heart, brain and other organs. Therefore, exposure to PM can result in serious impacts to health, especially in vulnerable groups of people such as the young, elderly, and those with respiratory problems. As a result, particulates are classified according to size. The UK is currently focused on measuring the fractions of PM where particles are less than 10 micrometres in diameter (PM10) and less than 2.5 micrometres in diameter (PM2.5) based on the latest evidence on the effects of PM to health. (UK Government 2024)

Table 4.1, from Shetty et al. (2023), is a brief and simple summary of pollutants, example sources, and health impacts on humans. Particulate matter are listed amongst other pollutants. While this is hardly a detailed discussion and exploration of our activities and their impact on the human body, it is however enough to make the point. Our negative space impact is not confined to our oceans and atmosphere.

Table 4.1 Major environmental pollutants, their sources, and impact on human health

Pollutants		Source	Impacts on human health
Heavy metals	Lead	Paints, Lead-acid batteries	Encephalopathy, Peripheral Neuropathy, Anaemia
	Mercury	Thermal power plants, hospital waste	Damage to the Liver, kidney, and brain, neurobehavioral changes, and abnormalities in fertility and pregnancy
	Arsenic	Wood preservatives, pesticide	Hypertension, Myocardial infarction, Proteinuria, cardiovascular diseases
	Nickel	Smelting operations, battery industries	Respiratory Cancer, Dermatomes, Genetic toxicity
	Cadmium	Tobacco smoke, batteries	Cancer, Dramatis
			Proteinuria, Glucosuria, Osteomalacia, Aminoaciduria, Emphysemia

Pollutants	Source	Impacts on human health
Sulphur dioxide (SO$_2$)	Fossil fuels combustion	Irritated airways and lungs. Prolonged exposure may lead to chronic bronchitis
Carbon monoxide (CO)	Vehicular emission, Open fire	Cardiovascular and pulmonary diseases, asphyxiation
Nitrogen oxides (NOx)	Fuel combustion	NOx gases can exacerbate respiratory illnesses and cardiovascular disease
Particulate matters	Vehicular emission	Chronic Pulmonary disease, bronchitis, asthma, respiratory and cardiovascular illness and mortality
PM2.5, PM10	Agricultural waste, Fuel, and wood burning	Stroke, change in blood pressure
Pesticides	Organochlorine compound	Dichloro-diphenyl-trichloroethane, DT, Dichlorodiphenyldichloroethane, Dicofol, Eldrin, Dieldrin

Damage human liver, kidney, neural and immune systems, and induces birth defects cancer, causes neurotoxicity, reproductive toxicity

(continued)

Table 4.1 (continued)

Pollutants	Source	Impacts on human health
Organophosphorus Compound	Malathion, parathion, diazinon, fenthion, dichlorvos, chlorpyrifos, ethion	Inflammation of the upper respiratory tract and bronchitis, blood effects such as aplastic anaemia Reproductive Effects Immunotoxicity
Carbamates	Sprays	Cancer and Immunosuppression Hypertension tachycardia, and paralysis Impair child development and IQ Decrease lung function Central nervous system tumour
Pyrethrin &Pyrethroids	Sprays, dust, and pet shampoos	Paranesthesia, respiratory tract, eyes, and skin irritations cardiovascular disease

Pollutants		Source	Impacts on human health
Plastics	High-density polyethylene	Plastic containers, pipes	Mild dermatitis, Respiratory damage, Hormone disruption
	Low-density polyethylene	Shrink wraps, squeeze bottles	Mild dermatitis, Burning sensation in eyes, Asthma
	Polyvinyl chloride	Cosmetic containers wrap	Respiratory damage, immune system damage
Plastic-Additives	Bisphenol A	Food storage containers,	Ovarian disorder
	Phthalates	Personal care products, Vinyl flooring, Polyvinyl chloride plastics	Endocrine disruptor Interference with testosterone, sperm motility, testicular cancer
	Dioxins	Tobacco smoke, Combustion of wood, coal, oil, Pesticides	Carcinogen interferes with testosterone
	Polycyclic aromatic hydrocarbons (PAHs)	Tobacco smoke, burning coal, oil, gas, wood, garbage	Developmental and reproductive toxicity
	Polychlorinated biphenyls (PCBs)	Contaminated fish, meat, and dairy products	Interferes with thyroid hormone

Source Verbatim from Shetty et al. (2023)

4.1.4   Continental Crust and Hydrosphere: Biodiversity Loss

Indeed, another primary example of our impact on space layers is biodiversity loss (IPBES 2019). I have used White et al.'s (2021) summary to present the extent of the biodiversity challenge before, I use it here again.

> Despite increasing recognition of its importance, biodiversity is in precipitous decline (Díaz et al. 2019; Tittensor et al. 2014). Recent reports estimate that 75% of the terrestrial environment and 66% of the marine environment have been severely altered by human activity (Halpern et al. 2015; IPBES 2019; Venter et al. 2016), and that between 1970 and 2014 populations of monitored species have declined by an average of 70% (WWF 2018). This decline is largely driven by the continued growth of the global economy (Hooke et al. 2012; IPBES 2019; Maxwell et al. 2016). From aquaculture and forestry to mining, consumer goods, and infrastructure, industrial development across sectors is closely tied to biodiversity loss. Business operations and supply chains act to increase the production and movement of goods, often at the expense of natural ecosystems through increasing habitat loss, fragmentation, pollution, invasive species introductions, and overexploitation (Díaz et al. 2019; Krausmann et al. 2017). (White et al. 2021)

4.1.5   Planetary Boundaries

Planetary Boundaries, based on the work done by Rockström et al. (2009), Steffen et al. (2015) and Richardson et al. (2023) at the Stockholm Resilience Centre, are not about specific space layers because they treat Earth as one system. I have decided to introduce them here to summarise our impact on Earth. The Planetary Boundaries framework is an important tool through which we can understand the critical levels of our destructive impact on our home planet. There is no contradiction between planetary boundaries and the space layers conceptualisation of our context. They each have very different purposes and objectives.

Following the work done at the Stockholm Resilience Centre, we now have evidence that out of 9 planetary boundaries we have already crossed 6, and a 7th, ocean acidification, is near threshold.

> [T]he planetary boundaries approach focuses on the biophysical processes of the Earth System that determine the self-regulating capacity of the planet. It incorporates the role of thresholds related to large-scale Earth System processes, the crossing of which may trigger non-linear changes in the functioning of

the Earth System, thereby challenging social–ecological resilience at regional to global scales. (Rockström et al. 2009, 5)

The nine planetary boundaries identified are Climate change, Biosphere integrity, Land System change, Freshwater change, Biochemical flows, Ocean acidification, Atmospheric aerosol loading, Stratospheric ozone depletion, and Novel entities. Figure 4.3 describes the historical record of measured and crossed boundaries (Richardson et al. 2023; SRC 2023). By 2023, we have crossed planetary boundaries with respect to climate change, biosphere integrity, novel entities, land system change, freshwater change, and biogeo-chemical flows. "This update on planetary boundaries clearly depicts a patient that is unwell" (SRC 2023).

4.1.6 Outer Space: Debris

Given that the focus and purpose of our discussion goes beyond Earth and aims to provide a snapshot of human impact across space layers, our physical context of matter stretching from subatomic to interstellar space and every layer in between and beyond, we must also look at our impact beyond Earth.

Our impact in outer space has not been any different in character. We have continuously contributed to the pollution of our outer space environment (see Fig. 4.4), leaving behind a trail of debris that pose risks to all missions.

> Today's unprecedented use of Earth's orbits coincides with increasingly unsus-tainable levels of space debris. Space debris already pose a direct collision risk to operational satellites and other spacecraft such as the International and Chinese space stations. This risk is expected to grow in the future, with planned projects numbering hundreds of thousands of satellites. (OECD 2024)

Indeed, on the 27th of June 2024, a day before the release of the above OECD report on the subject, Reuters reported that a "satellite has broken up into more than 100 pieces of debris in orbit, forcing astronauts on the International Space Station to take shelter for about an hour and adding to the mass of space junk already in orbit" (Reuters 2024).

While the number of tracked and catalogued debris is around 35,880, the European Space Agency (ESA) estimates that there are 40,500 debris objects greater than 10 cm, 1,100,000 debris objects from greater than 1 cm to 10 cm, and 130 million debris objects from greater than 1 mm to 1 cm. The total mass is estimated to exceed 12,400 tons (ESA 2024).

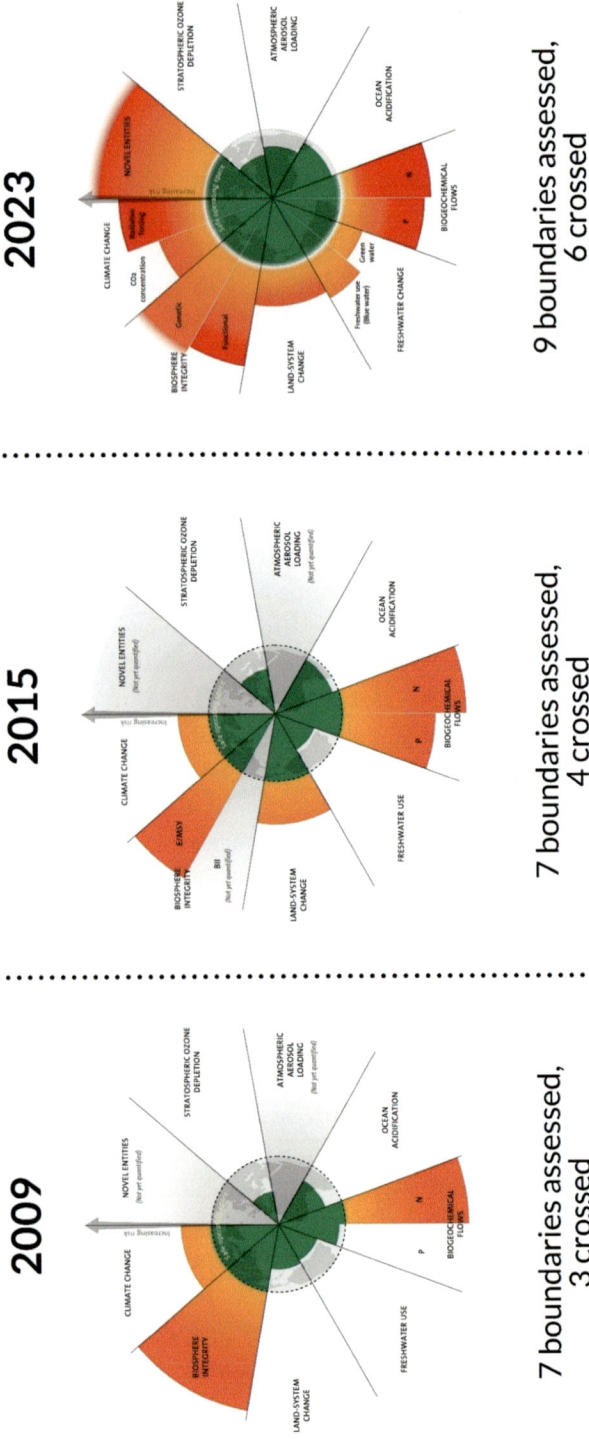

Fig. 4.3 Planetary boundaries (*Source* With permission from Stockholm Resilience Centre [2023])

Fig. 4.4 Outer space debris simulation >1 mm (*Source* ESA [2019])

4.2   Nihilistic Survival Condition

The previous sections provided a snapshot of our impact across many of the space layers depicted in Fig. 4.1. The picture is grim and reveals a human productivity that is ruthless and careless, oblivious to what it leaves behind and how it endangers its own evolutionary continuity. Can we expand into outer space when our space impact is as destructive as it is now? Should we reach out into the cosmos as a planet-consuming space-littering parasite?

As mentioned earlier and discussed in later chapters, I believe our current footprint and its destructive impact have a lot to do with our financial value framework, financial mathematics, and monetary architecture. In other words, with the appropriate theoretical, mathematical, and structural innovations we can address these issues and improve and optimise our space impact

But what if we were to continue down this current path, and not fix anything at all. What if we stayed on our current trajectory hoping that we could continue doing business as usual. Would we be able to expand into outer space and secure our sustainability and avoid self-inflicted extinction? In other words, can we secure our sustainability as a planet-consuming space-littering parasite?

Theoretically, this should be possible on the condition that we satisfy what I call the Nihilistic Survival Condition (NSC) or the Minimum Temporary Survival Condition for a Planet-Consuming Parasite (MTSCPCP). The condition is nihilistic because continuing on a course that is certain to destroy our current home implies discarding and rejecting the fundamental aspects of our existence that imply knowledge, morality, and/or meaning.

If we do not change trajectory and we continue destroying our ecosystem, the Nihilistic Survival Condition requires that humanity's rate of outer space expansion is at least double the rate of ecological destruction. In other words, *at the very least*, we must ensure that we are expanding twice as fast as we are destroying everything on our path.

Minimum Temporary Survival Condition for a Planet Consuming Parasite (MTSCAPCP)

$$\frac{\text{Rate of Outer Space Expansion}}{\text{Rate of Ecological Destruction}} = \frac{\text{ROSE}}{\text{RED}} = 2$$

$$\text{ROSE} = 2 \times \text{RED}$$

How can we measure these rates? While I have some ideas, I do not believe we should. We can and we must optimise our space impact across all layers of space.

4.3   Conclusion

This chapter discussed a select number of human impacts on the wider context of space, our physical context of matter stretching from subatomic to interstellar space and every layer in between and beyond, where outer space is but a segment. Naturally, other examples affecting other space layers can also be presented. Carbon in our atmosphere, plastic in our oceans, particulate matter in our lungs, debris in orbit, extinct species and deformed terrestrial environments, etc., are but some examples of our impact on space and its many layers.

We can find many more examples of waste and pollution. Municipal solid waste figures reveal the same trends. "More than two billion metric tons of municipal solid waste (MSW) are generated worldwide every year, and this figure is expected to increase by roughly 70 percent by 2050" (Statista 2023). Similarly, figures on sewage in our rivers and beaches, microplastics in our water and food chains, they all depict the same upward trend in environmental degradation and ecological destruction. Our footprint is careless and ruthless across all space layers.

The key conclusion is that human productivity is evidently oblivious to and unconcerned with how it affects the many layers of space it touches, and how it endangers its very own evolution and continuity. This, as I argue, is directly caused by our financial value framework, financial mathematics, and monetary architecture.

Indeed, as mentioned in the introduction, and discussed in Chapter 2, the same can be said about our inability to invest and expand our reach in outer space. They are both the consequence of our financial value framework, financial mathematics, and monetary architecture. I will explore and discuss these in more detail in the following chapters.

References

CIA. 2023. World Political Map. Central Intelligence Agency. Factbook. https://www.cia.gov/the-world-factbook/static/9a4e3054df7bfaf7101a6cefa28f2c8a/world_pol.jpg. Accessed 12 January 2024.

Copernicus. 2024. *June 2024 Marks 12th Month of Global Temperature Reaching 1.5°C Above Pre-industrial*. Copernicus

Díaz, S., et al. 2019. Pervasive Human-Driven Decline of Life on Earth Points to the Need for Transformative Change. *Science* 366(6471): eaax3100. https://doi.org/10.1126/science.aax3100. Accessed 2 February 2021.

ESA. 2019. Distribution of Space Debris in Orbit Around Earth. The European Space Agency. https://www.esa.int/ESA_Multimedia/Videos/2019/02/Distribution_of_space_debris_in_orbit_around_Earth. Accessed 12 April 2024.

EEA. 2023. How Air Pollution Affects Our Health. European Environment Agency. https://www.eea.europa.eu/en/topics/in-depth/air-pollution/eow-it-affects-our-health. Accessed 11 April 20204.

ESA. 2024. Space Debris by the Numbers. European Space Agency. https://www.esa.int/Space_Safety/Space_Debris/Space_debris_by_the_numbers. Accessed 15 July 2024.

Halpern, B.S., et al. 2015. Spatial and Temporal Changes in Cumulative Human Impacts on the World's Ocean. *Nature Communications* 6(1): 1–7. https://doi.org/10.1038/ncomms8615.

Hooke, R.L., Duque, M.J.F., Pedraza, J.D. 2012. Land Transformation by Humans: A Review. *GSA Today* 22: 4–10. https://www.geosociety.org/gsatoday/archive/22/12/pdf/i1052-5173-22-12-4.pdf. Accessed 2 February 2021.

IMF. 2023. IMF Fossil Fuel Subsidies Data: 2023 Update, International Monetary Fund. https://www.imf.org/en/Publications/WP/Issues/2023/08/22/IMF-Fossil-Fuel-Subsidies-Data-2023-Update-537281. Accessed 12 January 2024.

IOC-UNESCO. 2022. Ocean Plastic Pollution an Overview: Data and Statistics. https://oceanliteracy.unesco.org/plastic-pollution-ocean/. Accessed 2 February 2024.

IPBES. 2019. The Global Assessment Report on Biodiversity and Ecosystem Services. Intergovernmental Science-Policy Platform on Biodiversity and Ecosystem Services. https://ipbes.net/system/files/2021-06/2020%20IPBES%

20GLOBAL%20REPORT%28FIRST%20PART%29_V3_SINGLE.pdf. Accessed 2 February 2021.

IPCC. 2007. Climate Change 2007: The Physical Science Basis. The Intergovernmental Panel on Climate Change (IPCC). https://www.ipcc.ch/site/assets/uploads/2018/05/ar4_wg1_full_report-1.pdf. Accessed 12 December 2023.

IPCC. 2013. Climate Change 2013: The Physical Science Basis. Summary for Policymakers. Intergovernmental Panel on Climate Change. https://www.ipcc.ch/site/assets/uploads/2018/03/WG1AR5_SummaryVolume_FINAL.pdf. Accessed 2 February 2023.

IPCC. 2018. Summary for Policymakers. In Global Warming of 1.5 °C. IPCC. Available at https://www.ipcc.ch/site/assets/uploads/sites/2/2018/07/SR15_SPM_High_Res.pdf. Accessed 12 December 2023.

IPCC. 2021. Climate Change 2021: The Physical Science Basis. https://www.ipcc.ch/report/ar6/wg1/downloads/report/IPCC_AR6_WGI_SPM_final.pdf. Accessed 2 February 2022.

IPCC. 2022. Climate Change 2022: Impacts, Adaptation and Vulnerability. Summary for Policymakers. Intergovernmental Panel on Climate Change. https://report.ipcc.ch/ar6wg2/pdf/IPCC_AR6_WGII_SummaryForPolicymakers.pdf. Accessed 28 February 2022.

IPCC. 2023. Synthesis Report of the IPCC 6th Assessment Report (AR6): Summary for Policymakers. Intergovernmental Panel on Climate Change. https://report.ipcc.ch/ar6syr/pdf/IPCC_AR6_SYR_SPM.pdfreport.ipcc.ch/ar6wg2/pdf/IPCC_AR6_WGII_SummaryForPolicymakers.pdf. Accessed 22 march 2023.

Krausmann, F., et al. 2017. Global Socioeconomic Material Stocks Rise 23-Fold Over the 20th Century and Require Half of Annual Resource Use. Proceedings of the National Academy of Sciences of the United States of America 114 (8): 1880–1885. https://doi.org/10.1073/pnas.1613773114

Lebreton, L.C.M. et al. 2018. Evidence That the Great Pacific Garbage Patch Is Rapidly Accumulating Plastic. *Scientific Reports* 8: 4666. https://doi.org/10.1038/s41598-018-22939-w. Accessed 12 April 2023.

Maxwell, S.L., Fuller, R.A., Brooks, T.M., Watson, J.E. 2016. Biodiversity: The Ravages of Guns, Nets and Bulldozers. *Nature News* 536(7615): 143– 145. https://doi.org/10.1038/536143a.

NOAA. 2024. Trends in Atmospheric Carbon Dioxide. NOAA Global Monitoring Laboratory. https://gml.noaa.gov/ccgg/trends/. Accessed 1 February 2024.

Ocean Cleanup. 2024. What Is the great Pacific Garbage Patch? The Ocean Cleanup. https://theoceancleanup.com/great-pacific-garbage-patch/. Accessed 10 April 2024.

OECD. 2024. *The Economics of Space Sustainability: Delivering Economic Evidence to Guide Government Action.* Paris: OECD Publishing. https://doi.org/10.1787/b2257346-en. Accessed 2 July 2024.

Papazian, A. 2022. *The Space Value of Money: Rethinking Finance Beyond Risk and Time*. New York: Palgrave Macmillan. https://doi.org/10.1057/978-1-137-594 89-1.

Papazian, A. 2023. *Hardwiring Sustainability into Financial Mathematics: Implications for Money Mechanics*. New York: Palgrave Macmillan. https://doi.org/10.1007/978-3-031-45689-3.

PEW. 2020. Breaking the Plastic Wave. Pew Charitable Trusts. https://www.pewtru sts.org/-/media/assets/2020/10/breakingtheplasticwave_mainreport.pdf. Accessed 25 November 2023.

Reuters. 2024. Russian Satellite Breaks Up in Space, Forces ISS Astronauts to Shelter. https://www.reuters.com/technology/space/russian-satellite-blasts-debris-space-forces-iss-astronauts-shelter-2024-06-27/. Accessed 28 June 2024.

Richardson, K., et al. 2023. Earth Beyond Six of Nine Planetary Boundaries. *Science Advances* 9(37). https://doi.org/10.1126/sciadv.adh2458. Accessed 12 May 2024.

Rockström, J. et al. 2009. Planetary Boundaries: Exploring the Safe Operating Space for Humanity. *Ecology & Sociology* 14(2). https://www.ecologyandsociety.org/vol14/iss2/art32/. Accessed 12 March 2024.

Sagan, C. 1994. Pale Blue Dot. Random House.

Shetty, S.S., Deepthi, D., Harshitha, S., Sonkusare, S., Naik, P.B., Kumari, N.S., Madhyastha, H. 2023. Environmental Pollutants and Their Effects on Human Health. *Heliyon* 9(9):e19496. https://doi.org/10.1016/j.heliyon.2023.e19496. PMID: 37662771; PMCID: PMC10472068 (Aug 25).

SRC. 2023. All Planetary Boundaries Mapped Out for the First Time, Six of Nine Crossed. Stockholm Resilience Centre. https://www.stockholmresilience.org/res earch/research-news/2023-09-13-all-planetary-boundaries-mapped-out-for-the-first-time-six-of-nine-crossed.html. Accessed 12 May 2024.

Steffen, W., et al. 2015. Planetary Boundaries: Guiding Human Development on a Changing Planet. *Science* 347(6223). https://doi.org/10.1126/science.1259855. Accessed 12 February 2024.

Statista. 2023. Global Waste Generation—Statistics & Facts. https://www.sta tista.com/topics/4983/waste-generation-worldwide/#topicOverview. Accessed 12 January 2024.

Stockholm Resilience Centre. 2023. Planetary Boundaries over Time 2009, 2015, 2023. Stockholm Resilience Centre. Under the Creative Commons Licensing. https://stockholmuniversity.app.box.com/s/sr0nfknm95oydnnsm1zj0 c526qzjn1vs/folder/239674578598. Accessed 28 May 2024.

Tittensor, D.P., et al. 2014. A Mid-term Analysis of Progress Toward International Biodiversity *Targets*. *Science* 346(6206): 241–244. https://doi.org/10.1126/science.1257484.

UK Government. 2023. Hundreds of New North Sea Oil and Gas Licences to Boost British Energy Independence and Grow the Economy. https://www.gov.uk/government/news/hundreds-of-new-north-sea-oil-and-gas-licences-to-boost-british-energy-independence-and-grow-the-economy-31-july-2023. Accessed 4 November 2024.

UK Government. 2024. Particulate Matter (PM10/PM2.5). UK Government. https://www.gov.uk/government/statistics/air-quality-statistics/concentrations-of-particulate-matter-pm10-and-pm25. Accessed 11 May 2024.

UNFCCC. 2015. Paris Agreement. United Nations Framework Convention on Climate Change. https://unfccc.int/sites/default/files/english_paris_agreement.pdf. Accessed 2 December 2023.

Venter, O., et al. 2016. Sixteen Years of Change in the Global Terrestrial Human Footprint and Implications for Biodiversity Conservation. *Nature Communications* 7(1): 1–11. https://doi.org/10.1038/ncomms12558.

White, B.T., Viana, L.R., Campbell, G., Elverum, C., Bennun, L.A. 2021. Using Technology to Improve the Management of Development Impacts on Biodiversity. *Business Strategy and the Environment* 30: 3502–3516. https://doi.org/10.1002/bse.2816.

WWF. 2018. Living Planet Report—2018: Aiming Higher, ed. M. Grooten and R.E.A. Almond. WWF. https://www.worldwildlife.org/pages/living-planet-report-2018.

Part II

The Chains of Risk and Time

This part of the book explores the shortcomings of our current financial value framework, financial mathematics, and monetary architecture. It explains why and how they can be held responsible for our inability to invest and build sustainably on Earth, in space, and our inability to invest and expand in outer space.

5

Spaceless Financial Value Framework

Houston, we've had a problem.
James A. Lovell, Apollo 8 and 13 Astronaut, 1970

I'm not going into infinity. I'm going into low earth orbit.
Helen Sharman, Soyuz TM-12 Astronaut, 1991

Our inability to invest and expand our reach in outer space and our inability to invest and build a sustainable and fair reality on Earth, in space, are intricately related. They are both directly linked and caused by our financial value framework, financial mathematics, and monetary architecture. In this and following chapters I elaborate this argument and explain exactly how and why this is the case.

Our discussion must begin by taking a closer look at our financial value framework. The principles that define what and how we value are key to this framework, and they underpin the equations we have developed to measure value and return in finance. In this chapter, I discuss the principles of finance, their key propositions, and their implications.

To appropriately introduce our financial value framework, we must go back to 1776. Adam Smith, considered the founder of modern economics, in *The Wealth of Nations*, describes the motive and goal of investors or 'employers of capital' and how they interact with the interests of society as follows:

But it is only for the sake of profit that any man employs a capital in the support of industry; and he will always, therefore, endeavour to employ it in the support of that industry of which the produce is likely to be of the greatest

A. V. Papazian, *Financing the Race to Space*,
https://doi.org/10.1007/978-3-031-73102-0_5

value, or to exchange for the greatest quantity, either of money or of other goods.... He generally, indeed, neither intends to promote the public interest, nor knows how much he is promoting it... by directing that industry in such a manner as its produce may be of the greatest value, he intends only his own gain, and he is in this, as in many other cases, led by an invisible hand to promote an end which was no part of his intention... By pursuing his own interest he frequently promotes that of the society more effectually than when he really intends to promote it. (Smith 1776, 477)

The above much debated paragraph can be considered the seed that eventually mutated into our current framework. I say 'mutated' because our current framework is far more abstract than what Adam Smith initially presented. Nevertheless, the prominence and prioritisation of profit and the externalisation of responsibility are clear to see. Indeed, Milton Friedman's doctrine (Friedman 1970), "the social responsibility of business is to increase its profits," or the rationale of shareholder value maximisation, is a logical continuation of the same theoretical heritage. A self-serving rationality that seeks the greatest possible profit and value, in terms of money and/or other goods, that serves the greater good best by not actually serving it, all thanks to an invisible hand, is the precursor of our current financial value framework.

This chapter will bring the reader to the uncomfortable realisation that our financial value framework omits space. Space, as analytical dimension and as our physical context of matter, irrespective of constitution, composition, density, dynamics, and temperature, and stretching from subatomic space to interstellar space and every layer in between and beyond, is absent from our framework.

Indeed, the analytical framework upon which our models are built is spaceless. Our financial value framework, in academia and industry, in theory and practice, is built entirely on risk and time. It is designed to serve the investor, without any consideration of context, space, or impact. Moreover, and what is most important, this abstraction of space has also led to the omission of our responsibility for impact on space. Furthermore, as the discussion will reveal, the principles that govern our relationship with risk and time instil prohibitive biases against our evolutionary investments.

5.1   The Risk and Time Value of Money

As of today, the analytical framework of the finance discipline, in industry and academia, rests on two core principles: Risk and Return and Time Value of Money. This focus reveals a value paradigm built and defined by Risk and Time parameters, without any formal consideration of Space as an analytical dimension, nor any assessment of impact on space as our physical context. A review of finance literature in industry and academia supports this observation. (Papazian 2022, 17)

The two core principles that define the analytical framework of finance are Risk and Return and Time Value of Money. Both principles serve the investor and reveal specific features of this economic agent. Echoing the above introduction, the investor seeks to maximise returns, or profits, over time. The additional dimension of risk, which is the probability of a loss, describes a risk-averse investor who aims to minimise risks while maximising returns. In other words, the main and only stakeholder of our financial value framework is the mortal risk-averse return-maximising investor. Table 5.1 summarises the two principles.[1]

The principles in Table 5.1 define our analytical value framework in finance and are used to assess the value of cash flows. Cash flows, which represent a more refined conceptualisation of profits, are the amounts of cash flowing in and out of a venture or company or project. They are the ultimate purpose of any investment. While all cash flows are important, such as operating, investing, and financing cash flows, investors are mainly concerned with Free Cash Flows (FCF), the amount of net cash after taxes, and after all working capital and capital expenditure have been subtracted. The Corporate

---

[1] See also Papazian (2023, 2022), Williams (1938), Fama and French (1996, 2004, 2015), Gordon (1959), Gordon and Gordon (1997), Gordon and Shapiro (1956), Markowitz, (1952), Modigliani and Miller (1958), Ross (1976), Roll and Ross (1980), Sharpe (1964), Lintner (1965), Merton (1973), Black and Scholes (1973), Nobel Prize (1997), Koller et al. (2015, 2011), Choudhry (2012, 2018), Damodaran (2012, 2017), Yescombe (2014), Rosenbaum and Pearl (2013), Isaac and O'Leary (2013) and others for more evidence on the absence of space as analytical dimension and the focus on risk and time through the defined principles—risk and return and time value of money. The risk and time focus of the finance discipline is also revealed through the vast literature on stock market predictability, market efficiency, random walks, and overreaction (Papazian 2022; Fama 1970; Fama and French 1992, 1993; Malkiel 1973; De Bondt and Thaler 1985; Dissanaike, 1994,1997; Harvey et al. 2016; Xi et al. 2022; and others). See Brealey et al. (2020), Pike et al. (2018), and Watson and Head (2016) for popular textbooks in the field. See Haynes (1895), Knight (1921), and Markowitz (1952) for a discussion on risk.

Table 5.1 The core principles of finance theory and practice

Stakeholder	Risk	Time
Mortal Risk-Averse Return-maximising Investor	**Risk and Return**: The higher the risk the higher the expected return— given the risk-averse nature of investors, higher risks imply higher expectations of reward	**Time Value of Money**: A dollar ($1) today is worth more than a dollar ($1) tomorrow—because a dollar today can earn interest/return by tomorrow and be more than a dollar by tomorrow

(*Source* Papazian [2023])

Finance Institute defines Free Cash Flow as follows (CFI 2024)[2]:

$$FCF = Net\,Income + Non\,Cash\,Expenses - Increase\,in\,Working\,Capital - Capital\,Expenditures$$

The time value of money principle (a dollar today is worth more than a dollar tomorrow), implies that the more distant in time the cash flow is, the less it is worth today. In other words, when assessing the expected Free Cash Flows, investors will ascribe less value when cash flows and returns are further away in time. Similarly, the risk and return principle (the higher the risk the higher the expected return), implies that the riskier the cash flow, the higher the expected return by investors, and therefore the cost of capital. Brealey et al. (2020) in their 13th edition finance textbook called 'Principles of Corporate Finance,' state the following in their introduction: "a safe dollar is worth more than a risky dollar" (Brealey et al. 2020, 1). In other words, investors would expect and require higher returns the higher the risk of an investment, and thus they would discount risky dollars with a higher cost of capital.

Based on the above, the only stakeholder of these principles is the mortal risk-averse return-maximising investor whose optimisation target is to maximise returns while minimising risks. I have added the adjective 'mortal'

---

[2] Non-cash expenses are added given that they do not imply any actual outflow of cash. Operating cash flows represent the cash amounts that a company's operating activities generate (inflow) or consume (outflow). Investing cash flows represent the cash amounts that a company generates (inflow) or consumes (outflow) in its investment activities. Financing cash flows represent the amounts of cash a company raises or borrows to fund the company. Working capital is the difference between a company's liquid assets and liabilities (Current) and is used to fund operations and meet financial requirements. Capital expenditures (CapEx) are the cash expenditures that are dedicated to the acquisition and maintenance of physical assets like property, equipment, and technology.

(Papazian 2022, 2023) because it emphasises the mindset and nature of the investor. This is because risk and time are very mortal concerns. Indeed, an immortal investor would hardly be concerned with time and/or risk, as it would be ever present in space. By contrast to the mortal individual in the chain, the human collective, which can procreate and secure its continuous existence, can be considered eternal through procreation. Given this more complex equation, an eternal society made up of mortal individuals, the human collective would be more concerned with its evolutionary continuity rather than individual risk and time concerns (Papazian 2023, 2022).

The point here is that our financial value framework is built on two principles of value that assess the value of cash flows based on risk and time, without space. Our physical context and the impact of cash flows on our physical context are left out. Our core analytical framework in finance is spaceless.

The absence of space, as analytical dimension and our physical context, from our financial value framework is a serious omission that explains our current predicament. But it is not the only issue, and it is only the most abstract level of our challenge.

5.2   Risk, Time, and Our Evolutionary Investments

Another key and debilitating implication of our financial value framework, besides the omission of space, is that the existing principles of value in finance, Risk and Return and Time Value of Money, in their very logic, discriminate against our evolutionary investments.

> All our evolutionary challenges require investments in the present that carry *very high risks* and *distant returns* - features that are negatively priced based on the current principles of value that underpin the equations we use and teach in the field.

> The current principles of value in finance theory leave our evolutionary investments in a blind spot. By negatively pricing distant returns and high risks, our financial value framework misprices our evolutionary investments. In fact, today, our evolutionary investments become plausible and 'affordable' only when they can be made to make sense within the preference framework of the mortal risk-averse return-maximising investor. A theoretical and practical misconception that could well explain our current predicament. (Papazian 2023, 14)

Outer space exploration, development, and settlement is an evolutionary challenge that unavoidably involves very high risks and very distant returns—features that are negatively priced thanks to the core principles of finance.

Indeed, given this in-built bias against risky and distant dollars, we should not be surprised with the figures presented in Chapter 2. An Earthbound private outer space economy chasing Earthly money supply is, in truth and without a doubt, serving and satisfying this framework. Investors, as per the principles that govern the entire discipline and industry of finance, are chasing cash flows—and the less risky and less distant they are the better.

This is why Elon Musk's approach, an evolutionary vision and an openly declared mission to make humanity a multi-habitat species, is not just unusual, it also challenges the financial framework SpaceX must survive and grow in (Musk, 2017a,b). Revisiting Elon's tweet referenced in Chapter 2,[3] we can see evidence of this challenge. Humanity's future in outer space must be paid for through the revenues of Starlink, an internet service provider, because our financial value framework does not attach any particular value to the evolutionary vision and mission.

In truth, and in parallel, we should also not be surprised with humanity's destructive impact on space. When our core principles of finance omit our context, i.e., space, when our principles do not establish any responsibility whatsoever, our careless and brazen disrespect for space is understandable. Responsibility for space impact is abstracted away, and it is considered outside the decision framework that governs our investments. This is why corporate social responsibility and sustainability are treated outside our core equations and models in finance. They are optional addendums rather than structural components.

The guiding system of our investments and our productive activities seems to be aimed at serving an individual mortal in the chain, rather than the human collective. Indeed, our financial value framework serves the mortal risk-averse return-maximising investor and her/his/their preferences and has inbuilt biases against our evolutionary investments with high risks and distant returns.

The continuous destruction of our ecosystem and our inability to invest and change course are all anchored in the value framework that underpins this guiding system—our financial value framework.

---

[3] "SpaceX might exceed 90% of all Earth payload to orbit later this year. Once Starship is launching at high rate, probably >99%. Has to be or we can't build a city on Mars or base on moon. We file almost no patents, so nothing stopping competition from copying us" (Musk 2024).

5.3   Conclusion

Given that our current financial value framework is built around Risk and Return and Time Value of Money, the evaluation of profits or value and return is based solely on risk and time parameters, without any context or space parameters. The only stakeholder formally considered in our framework is the mortal risk-averse return-maximising investor whose optimisation target is the maximisation of return and the minimisation of risk. The abstraction of space from our framework leads to the omission of our responsibility for space impact. In other words, the mortal risk-averse return-maximising investor is absolved of all the potential and actual damage that investments and cash flows inflict in and on space.

The principles upon which the discipline and industry of finance are built discriminate against our evolutionary investments thanks to their very logic. This is so because all our evolutionary challenges and investments imply distant cash flows and very high risks and both features are negatively priced—a safe dollar is worth more than a risky dollar, and a dollar today is worth more than a dollar tomorrow. These are 'innocent' assumptions that leave our evolutionary investments in a blind spot. In fact, they lead to the mispricing of our evolutionary investments.

When our principles define the value of cash flows in relation to risk and time alone and discriminate against our evolutionary investments, allocating resources to highly risky and long-horizon endeavours is an institutional, theoretical, and mathematical challenge for finance theory and practice.

Omitting our physical context of matter, i.e., space, from the core framework that defines the guiding system of our investments may be a theoretical slip up, but it has profound implications on our ecosystem and our evolutionary continuity. Indeed, our productive capacity and civilisation are chained to risk and time, and much of our destructive and constrained footprint in outer space is a function of these chains and the incentives and requirements they impose.

The above challenges presented by our core principles of value in finance are critical but only one part of the theoretical fog that has led to the destruction of our ecosystem, as well as the neglect of space, outer space, and our key evolutionary investments. As the next chapter will reveal, this spaceless financial value framework has resulted in a space-blind financial mathematics.

References

Black, F., Scholes, M. 1973. The Pricing of Options and Corporate Liabilties. *The Journal of Political Economy* 81: 637–654. https://www.jstor.org/stable/1831029. Accessed 2 February 2023.

Brealey, A.R., Myers, C.S., Allen, F. 2020 *Principles of Corporate Finance*. 13th Ed. New York: McGraw Hill.

Choudhry, M. 2012. *The Principles of Banking*. Singapore: Wiley.

Choudhry, M. 2018. *Past, Present, and Future Principles of Banking and Finance*. Singapore: Wiley.

CFI. 2024. *Free Cash Flow (FCF) Formula*. Corporate Finance Institute. https://corporatefinanceinstitute.com/resources/valuation/fcf-formula-free-cash-flow/. Accessed 12 May 2024.

Damodaran, A. 2012. *Investment Valuation*. 3rd Ed. New Jersey: Wiley.

Damodaran, A. 2017. *Damodaran on Valuation*. 2nd Ed. New Jersey: Wiley.

De Bondt, W.F.M., Thaler, R. 1985. Does the Stock Market Overreact? *The Journal of Finance* 40 (3): 893–805. https://doi.org/10.2307/2327804. Accessed 2 February 2021.

Dissanaike, G. 1997. Do Stock Market Investors Overreact? *Journal of Business Finance and Accounting* 24(1): 27–50. https://doi.org/10.1111/1468-5957.00093. Accessed 2 February 2021.

Dissanaike, G. 1994. On the Computation of Returns in Tests of the Stock Market Overreaction Hypothesis. *Journal of Banking & Finance* 18(6): 1083–1094. https://doi.org/10.1016/0378-4266(94)00061-1. Accessed 2 February 2022.

Fama, E.F. 1970. Efficient Capital Markets: A Review of Theory and Empirical Work. *The Journal of Finance* 25: 383–417.

Fama, E.F., French, K.R. 1992. The Cross-Section of Expected Stock Returns. *The Journal of Finance* 47: 427–465. https://doi.org/10.2307/2329112. Accessed 2 February 2021

Fama, E.F., French, K.R. 1993. Common Risk Factors in the Returns on Stocks and Bonds. *The Journal of Financial Economics* 33: 3–56. https://doi.org/10.1016/0304-405X(93)90023-5. Accessed 2 February 2021.

Fama, E.F., French, K.R. 1996. Multifactor Explanations of Asset Pricing Anomalies. *The Journal of Finance* 51: 55–84. https://doi.org/10.1111/j.1540-6261.1996.tb05202.x. Accessed 2 February 2021

Fama, E.F., French, K.R. 2004. The Capital Asset Pricing Model: Theory and Evidence. *Journal of Economic Perspectives* 18: 25–46. https://doi.org/10.1257/0895330042162430. Accessed 2 February 2022.

Fama, E.F. French, K.R. 2015. A Five-Factor Asset Pricing Model. *Journal of Financial Economics* 116: 1–22. https://doi.org/10.1016/j.jfineco.2014.10.010. Accessed 2 February 2021.

Friedman, M. 1970. A Friedman Doctrine—The Social Responsibility of Business is to Increase Its Profits. *The New York Times*. https://www.nytimes.com/1970/

13/archives/a-friedman-doctrine-the-social-responsibility-of-business-is-to.html. Accessed 8 November 2023.

Gordon, J.R., Gordon, M.J. 1997. The Finite Horizon Expected Return Model. *Financial Analysts Journal* 53: 52–61. https://doi.org/10.2469/faj.v53.n3.2084. Accessed 2 February 2021.

Gordon, M.J., Shapiro, E. 1956. Capital Equipment Analysis: The Required Rate of Profit. *Management Science* 3: 102–110. https://www.jstor.org/stable/2627177. Accessed 2 February 2021.

Gordon, M.J. 1959. Dividends, Earnings, and Stock Prices. *The Review of Economics and Statistics* 41: 99–105. https://doi.org/10.2307/1927792. Accessed 2 February 2021.

Harvey, R., Liu, Y., Zhu, H. 2016. ... and the Cross-Section of Expected Returns. *The Review of Financial Studies* 29: 5–68. https://doi.org/10.1093/rfs/hhv059. Accessed 2 February 2021.

Isaac, D., O'Leary, J., 2013. *Property Valuation Techniques*. 3rd Ed. London: Palgrave Macmillan.

Haynes, J. 1895. Risk as an Economic Factor. *The Quarterly Journal of Economics* 9(4): 409–449. https://doi.org/10.2307/1886012. Accessed 2 February 2022.

Knight, F.H. 1921. *Risk, Uncertainty and Profit*. Houghton Mifflin Company.

Koller, T., Goedhart, M., Wessels, D., McKinsey and Company. 2015. *Valuation: Measuring and Managing the Value of Companies*. 6th Ed. New Jersey: Wiley.

Koller, T., Dobbs, R., Huyett, B., McKinsey and Company. 2011. *Value: The Four Cornerstones of Corporate Finance*. 6th Ed. New Jersey: Wiley.

Lintner, J. 1965. The Valuation of Risk Assets and the Selection of Risky Investments in Stock Portfolios and Capital Budgets. *The Review of Economics and Statistics* 47: 13–37. https://doi.org/10.2307/1924119. Accessed 2 February 2023.

Malkiel, B.G. 1973. *A Random Walk Down Wall Street*. New York: W. W. Norton.

Markowitz, H. 1952. Portfolio Selection. *The Journal of Finance* 7: 77–91. https://doi.org/10.2307/2975974. Accessed 2 February 2023.

Merton, R. 1973. An Intertemporal Capital Asset Pricing Model. *Econometrica* 41: 867–887. https://doi.org/10.2307/1913811. Accessed 2 February 2023.

Modigliani, F., Miller, M.H. 1958. The Cost of Capital, Corporation Finance and the Theory of Investment. *The American Economic Review* 48: 261–297. https://www.jstor.org/stable/1809766. Accessed 2 February 2023.

Musk, E. 2024. Tweet on Q1 2024 Orbital Launch Report. https://x.com/elonmusk/status/1792689139470704718. Accessed 20 May 2024.

Musk, E. 2017a. Making Life Multiplanetary. SpaceX. https://www.spacex.com/media/making_life_multiplanetary_transcript_2017.pdf. Accessed 12 March 2024.

Musk, E. 2017b. Making Humans a Multi-Planetary Species. New Space. https://doi.org/10.1089/space.2017.29009.emu. Accessed 12 March 2023.

Nobel Prize. 1997. For a new method to determine the value of derivatives. The Nobel Prize. Press Release. https://www.nobelprize.org/prizes/economic-sciences/1997/press-release/. Accessed 02 February 2022.

Papazian, A. 2022. *The Space Value of Money: Rethinking Finance Beyond Risk and Time*. New York: Palgrave Macmillan. https://doi.org/10.1057/978-1-137-594 89-1.

Papazian, A. 2023. *Hardwiring Sustainability into Financial Mathematics: Implications for Money Mechanics*. New York: Palgrave Macmillan. https://doi.org/10.1007/978-3-031-45689-3.

Pike, R., Neale, B., Akbar, S., Linslley, P. 2018. *Corporate Finance and Investment*. 9th Ed. London: Pearson.

Roll, R., Ross, S.A. 1980. An Empirical Investigation of the Arbitrage Pricing Theory. *The Journal of Finance* 35: 1073–1103.

Rosenbaum, J., Pearl, J. 2013. *Investment Banking*. New Jersey: Wiley.

Ross, S.A. 1976. The Arbitrage Theory of Capital Asset Pricing. *Journal of Economic Theory* 13: 341–360. https://doi.org/10.1016/0022-0531(76)90046-6. Accessed 2 February 2023.

Sharpe, W.F. 1964. Capital Asset Prices: A Theory of Market Equilibrium Under Conditions of Risk. *Journal of Finance* 19: 425–442. https://doi.org/10.1111/j.1540-6261.1964.tb02865.x. Accessed 2 February 2023.

Smith, A. 1776. *The Wealth of Nations*. Cannan, E. 1961. London: Methuen.

Watson, D., Head, A. 2016. *Corporate Finance: Principles and Practice*. 7th ed. Pearson, London.

Williams, J.B. 1938. *The Theory of Investment Value*. Cambridge: Harvard University Press.

Xi, D., Yan, L., Rapach, D.E., Zhou, G. 2022. Anomalies and the Expected Market Return. *The Journal of Finance* 77: 639–681. https://doi.org/10.1111/jofi.13099. Accessed 20 February 2022.

Yescombe, E.R. 2014. *Principles of Project Finance*. 2nd Ed. Oxford: Academic Press.

6

Spaceless Financial Mathematics

If we can conquer space, we can conquer childhood hunger.
Buzz Aldrin, Gemini 12 and Apollo 11 Astronaut, 1966

Only when I saw the Earth from space, in all its ineffable beauty and fragility, did I realise that humankind's most urgent task is to cherish and preserve it for future generations.
Sigmund Jähn, Soyuz 31 Cosmonaut, 1978

A spaceless financial value framework has led us to a space blind financial mathematics. The abstraction of space, as analytical dimension and our physical context, is reflected in the entire corpus of equations that define and theorise value and return in finance. They are all based on risk and time parameters, gauging the value of cash flows without any space parameters. Focused on measuring the risk and time value of money, our equations do not consider the space impact of cash flows as relevant.

The omission of space impact from our core equations is a structural impediment that explains both, our inability to invest and build sustainably on Earth, and our inability to invest and expand our reach in outer space. The first is explained by the fact that the abstraction of space has led to the omission of space impact from our equations, and the complete absence of our responsibility for impact on space. The second is explained by the fact that outer space exploration and settlement projects are endowed with abundant space impact, have high risks, and distant returns. While the latter two

A. V. Papazian, *Financing the Race to Space*,
https://doi.org/10.1007/978-3-031-73102-0_6

are negatively priced by our principles, the former is omitted. Our equations, therefore, are structurally biased against outer space investments.

The purpose of this chapter is to demonstrate the absence of space and space impact from our financial mathematics, specifically our equations of value and return in finance.

6.1   Equations Without Space, Without Space Impact

In Table 6.1 you can find a sample of bond, stock, asset, firm, option, and cash flow valuation equations commonly used in finance. Some of them have earned their inventors the coveted Nobel Prize in economics. The equations reveal a financial mathematics without space, or outer space. There are no context parameters in these equations, and they are all structured around risk and time proxies, assessing the value of cash flows, assets, and instruments in risktime.

Looking closely at the equations and their elements you will notice that time and risk parameters are used to measure the value and/or return of firms, cash flows, and instruments. The space impact that it would take to achieve the cash flows and returns is not relevant. Indeed, the core equations of finance, built on our financial value framework, endorsed by the discipline and applied by the industry, have institutionalised the omission of space and our responsibility for impact on space. Therefore, our current predicament and destructive footprint should not come as a surprise.

These equations define and influence where, how, and why we invest. Even when they are not applied literally and in isolation, they define the mindset of investors, and underpin our markets. This is perpetuated through financial education in academia and industry. When we teach students and analysts to maximise returns and minimise risks, without any consideration of space and their responsibility for impact, we entrench the framework and mathematics in the very fabric of our markets. Indeed, through the above omissions, finance absolves investors of their negative impact on space, and externalises our responsibility through optional addendums.

As I have argued in Papazian (2023, 2022), it really does not matter how many new climate-related and ESG reporting standards and frameworks we invent, none of them will have the desired impact when our financial value framework and financial mathematics are blind to space and omit our responsibility for impact from our core equations—taught and applied by millions around the world. The current standards and frameworks of sustainability,

Table 6.1 Sample finance equations: bonds, firms, stocks, assets, options

Sample bond valuation equations

$$\text{Bond Value} = \sum_{t=1}^{n} \frac{C_t}{(1+r)^t} + \frac{P}{(1+r)^t}$$

$$\text{Bond Value} = \sum_{t=1}^{n \times m} \frac{C_t}{(1+(\frac{r}{m}))^t} + \frac{P}{(1+(\frac{r}{m}))^{n \times m}}$$

Sample of stock and firm valuation equations

$$P_0 = \frac{D_1}{r-g}$$

$$P_0 = \sum_{t=1}^{n} \frac{D_t}{(1+\text{WACC})^t} + \frac{P_n}{(1+\text{WACC})^t}$$

$$P_0 = \sum_{t=1}^{n} \frac{D_t}{(1+\text{WACC})^t} + \frac{D_{n+1}}{(\text{WACC}-g) \cdot (1+\text{WACC})^n}$$

$$\text{Firm Value} = \sum_{t=1}^{n} \frac{\text{FCFF}_t}{(1+\text{WACC})^t} + \frac{\text{FCFF}_{n+1}}{(\text{WACC}-g) \cdot (1+\text{WACC})^n}$$

Sample of asset pricing models

$$R_i = R_f + \beta_i \times (R_m - R_f) \quad \beta_i = \frac{\text{Covariance}_{R_i, R_m}}{\text{Variance}_{R_m}}$$

Elements

n = Maturity or Number of Periods
m = Number of Compounding
P = Par Value of Bond
C_t = Coupon Payments

P_0 = Stock price
g = Constant Growth Rate in Dividends
r = Constant Cost of Capital
D_1 = Next Year/Period Dividend
P_n = Terminal Value = $(D_{n+1}/\text{WACC}-g)$
D_t = Dividend at t
D_{n+1} = Dividend at $n+1$
WACC = Weighted Average Cost of Capital
g = Constant Growth Rate in Dividends
FCFF_t = Free Cash Flow to Firm at t

R_i = Return on security i
R_f = Risk Free Rate
β_i = Beta = Systematic Risk Proxy
Rm = Return on market

(continued)

Table 6.1 (continued)

$E(R_i) - R_f = b_1(E(R_M) - R_f) + s_i E(\text{SMB}) + h_i E(\text{HML})$

$E(R_i) - R_f = $ Expected Excess Return Stock i
$E(R_i) = $ Expected Return on Stock i
$R_f = $ Risk Free Rate
$E(R_M) = $ Expected Return on Market
$E(R_M) - R_f = $ Exp. Market Risk Premium
$E(\text{SMB}) = $ Expected Size Premium
$E(\text{HML}) = $ Expected Value Premium
$b_i, s_i, h_i = $ Factor Sensitivities or Loadings

Modigliani Miller corporate value & capital structure model

$V_j = (S_j + D_j) = \frac{\overline{X_j}}{\rho_k}$

$V_j = $ Value of Firm j
$S_j = $ Market Value of Common Shares of j
$D_j = $ Market Value of Debts of j
$X_j = $ Expected Return on the Assets owned by the company
$\rho_k = $ Capitalisation Rate for shares in class k

$i_j = \rho_k + (\rho_k - r)\frac{D_j}{S_j}$

Black and Scholes option pricing model

$C = SN(d) - Le^{-rt}N(d - \sigma\sqrt{t})$

$C = $ Value of Call Option
$Nd = $ Normal Distribution Function
$t = $ Time to Maturity
$L = $ Exercise (Strike)Price of Option
$\sigma = $ Standard Deviation of Return on Stock
$r = $ Risk Free Interest Rate
$S = $ Current Stock Price or Asset Price

$d = \frac{\ln\frac{S}{L} + \left(r + \frac{\sigma^2}{2}\right)t}{\sigma\sqrt{t}}$

Net present value & cash flow valuation

$$\text{NPV} = \sum_{t=0}^{T} \frac{\text{CF}_t}{(1+r)^t} \qquad \text{NPV} = \text{CF}_0 + \sum_{t=1}^{T} \frac{\text{CF}_t}{(1+r)^t}$$

n = Time Horizon

t = Moving time

r = Discount Rate

II = Initial Investment

CF_t = Future Expected Cash Flows at t

$$\text{Net Present Value} = -II + \sum_{t=1}^{n} \frac{\text{CF}_t}{(1+r)^t}$$

You can find a detailed discussion in Papazian (2022). See Brealey et al. (2020), Pike et al. (2018), Watson and Head (2016), Williams (1938), Fama (1970), Fama and French (1992, 1993, 1996, 2004, 2015), Gordon (1959), Gordon and Gordon (1997), Gordon and Shapiro (1956), Markowitz (1952), Modigliani and Miller (1958, 1963), Ross (1976), Roll and Ross (1980), Sharpe (1964), Lintner (1965), Harvey et al. (2016), Merton (1973), Black and Scholes (1973), Nobel Prize (1997), Koller et al. (2015, 2011), Choudhry (2012, 2018), Damodaran (2012, 2017), Yescombe (2014), Rosenbaum and Pearl (2013), Isaac and O'Leary (2013) and others. See Papazian (2023, 2022) for a detailed discussion of the above, and the absence of space and space impact.

Source Adapted and updated from Papazian (2023)

while well-intentioned, are a sophisticated distraction. Indeed, they add burdens on business and industry and, by their very nature, cannot deliver the desired and promised outcome, i.e., a sustainable human civilisation.

Thus, the omission of space impact from our core equations undermines our ability to invest and build sustainably on Earth, in space. The absence of space impact from our equations is also a primary reason that explains our inability to invest and expand our reach in outer space. This is because outer space exploration and settlement projects are rich in space impact and have high risks and distant returns. While the first feature is missing from our equations, the other two are negatively priced thanks to our principles.

The guidance system of our investments is blind to space and omits our responsibility for impact. It also does not value our footprint, and/or its expansion. They are irrelevant and excluded from our equations. This leads to an environment where outer space projects and investments, directly and indirectly, are much harder to justify, finance, and support. Naturally, this is particularly relevant to the private outer space economy. The pursuit of Earthly money supply is grounded in this framework, defined by the minimisation of risks and maximisation of return.

Naturally, there are investors who have a larger appetite for risk, and they may be more patient as well. This discussion does not negate the presence of investors who choose to invest for space impact, taking on higher risks and distant return expectations. The issue is that the framework and mathematics that govern our markets and the vast majority of our investments do not consider space and our impact to be of inherent value.

Interestingly, a good proportion of our equations, namely those that use discounting as a method of valuation, suffer from yet another bias.

6.2 Discounting the Imaginary, Abstracting the Actual

As discussed, and demonstrated, structured around risk and time, serving the mortal risk-averse return-maximising investor, with inbuilt biases against our evolutionary investments, and an entire dimension of context missing, i.e., space and outer space, our financial value framework and equations omit space impact and do not equip us with the tools through which we can value our evolutionary investments in space. As of today, our financial value framework and mathematics misprice our evolutionary investments, like outer space development, exploration, and settlement projects.

There is more to this story. A good proportion of our equations, namely discounting models, reveal yet another key bias. Taking as an example the Net Present Value (NPV) equation, which epitomises a risktime conceptualisation of value in finance, we observe that the mathematical focus is on the non-actual future expected cash flows rather than the actual investment.

One of the most commonly used tools in finance textbooks is the cash flow timeline (see Fig. 6.1). The very logic and use of the cash flow timeline substantiates the above. The focus is on the future non-actual cash flows. This is further confirmed by the NPV equation (Eq. 6.1). The NPV equation[1] discounts future expected cash flows to the present using a discount rate (a proxy for the risk levels involved in the opportunity measured through the return of an alternative investment with the same level of risk) and subtracts the initial investment from the Present Value of the future cash flows. It has two parts, an actual part, the initial investment, and a non-actual part, the future expected cash flows.

$$\text{Net Present Value} = -II + \sum_{t=1}^{n} \frac{CF_t}{(1+r)^t} \tag{6.1}$$

$n = \text{Time Horizon}$

$t = \text{Moving time}$

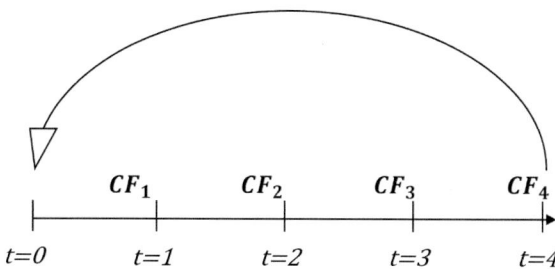

$CF_1 \quad CF_2 \quad CF_3 \quad CF_4$

$t=0 \quad t=1 \quad t=2 \quad t=3 \quad t=4$

Fig. 6.1 Future expected cash flow timeline (*Source* Author)

---

[1] The NPV equation is sometimes written in the below formats, where the first cashflow CF_0 (II) is included in the right-side term as the first cash flow at $t = 0$, or excluded but without the negative sign as the negative sign of the first cash flow CF_0 is assumed:

$$\text{NPV} = \sum_{t=0}^{T} \frac{CF_t}{(1+r)^t} \quad \text{NPV} = CF_0 + \sum_{t=1}^{T} \frac{CF_t}{(1+r)^t}$$

$r = $ Discount Rate

$II = $ Initial Investment

$CF_t = $ Future Expected Cash Flows

Future expected cash flows are non-actual or imaginary because they have not happened yet. They may happen as expected or agreed, or they may not. If these cash flows were guaranteed, there would be no need to discount them into the present to account for their riskiness over time. Naturally, applying a discount rate to the future expected cash flows does not make the cash flows any less non-actual, or more real. (Papazian, 2023, 2022)

The mathematical attention of our models is focused on the future *non-actual* element, the expected cash flows, and *not* on the *actual* element, the initial investment. The actual element of the equation gets minimal treatment. We observe that the initial investment (II) is treated with a '-' sign to denote an outflow for the mortal risk-averse return-maximising investor (Papazian, 2023, 2022).[2] Meanwhile, the future expected and non-actual cash flows are treated for risk and time. The only thing that matters is the fact that it is an outflow for the mortal risk-averse return-maximising investor. Thus, a negative sign is considered enough treatment for the initial investment (II).

Indeed, we cannot be surprised with the state of the world. We cannot and should not be surprised with the levels of funding allocated to outer space development and exploration—they require massive investments in the present, involve immense risks, promise only very distant cash flows, and much of their value is in their space impact. Meanwhile, our equations are focused on the non-actual future cash flows, omit space and space impact, and are built around risk and time, based on principles that attach greater value to safe dollars today.

---

[2] The Net Present Value equation (NPV) is one of the most commonly used equations. Graham and Harvey (2002) reveal that Net Present Value (NPV) is one of the most frequently used capital budgeting techniques by CFOs (Papazian 2022).

6.3   Conclusion

It is very hard to see how humanity can create and allocate the necessary amounts to build and maintain our expansion in outer space with a financial value framework and mathematics built in risktime, with principles that discriminate against our evolutionary investments, and equations that are more concerned with the non-actual future expected cash flows than their actual space impact.

In Chapter 2, when discussing the private outer space economy, we observed that it is Earth centric. This is a default scenario given how our framework and mathematics work. When outer space companies are expected to deliver free cash flows, and those expected cash flows are more important than the actual space impact of the projects, and they can only be earned or received through paying customers on Earth, the sector can only be Earthbound.

The only segment of the outer space economy that is not chained to Earth in this way is made up of the public agencies and programs. Governmental outer space agency expenditures do not expect a monetary return, and so they are not chained to the Earthly money supply when venturing out. However, even though not obligated nor expected to make a profit, this framework and mathematics chains the public sector as well. This so through government budgets, and the taxes and debts that fund them. This brings us to the next topic, our spaceless debt-based monetary architecture, which plays a central role in constraining our ability to expand our reach in outer space.

References

Black, F., Scholes, M. 1973. The Pricing of Options and Corporate Liabilties. *The Journal of Political Economy* 81: 637–654. https://www.jstor.org/stable/1831029. Accessed 2 February 2023.

Brealey, A.R., Myers, C.S., Allen, F. 2020. *Principles of Corporate Finance*. 13th Ed. New York: McGraw Hill.

Choudhry, M. 2012. *The Principles of Banking*. Singapore: Wiley.

Choudhry, M. 2018. *Past, Present, and Future Principles of Banking and Finance*. Singapore: Wiley.

Damodaran, A. 2012. *Investment Valuation*. 3rd Ed. New Jersey: Wiley.

Damodaran, A. 2017. *Damodaran on Valuation*. 2nd Ed. New Jersey: Wiley.

Fama, E.F. 1970. Efficient Capital Markets: A Review of Theory and Empirical Work. *The Journal of Finance* 25: 383–417.

Fama, E.F., French, K.R. 1992. The Cross-Section of Expected Stock Returns. *The Journal of Finance* 47: 427–465. https://doi.org/10.2307/2329112. Accessed 2 February 2021.

Fama, E.F., French, K.R. 1993. Common Risk Factors in the Returns on Stocks and Bonds. *The Journal of Financial Economics* 33: 3–56. https://doi.org/10.1016/0304-405X(93)90023-5. Accessed 2 February 2021.

Fama, E.F., French, K.R. 1996. Multifactor Explanations of Asset Pricing Anomalies. *The Journal of Finance* 51: 55–84. https://doi.org/10.1111/j.1540-6261.1996.tb05202.x. Accessed 2 February 2021

Fama, E.F., French, K.R. 2004. The Capital Asset Pricing Model: Theory and Evidence. *Journal of Economic Perspectives* 18: 25–46. https://doi.org/10.1257/0895330042162430. Accessed 2 February 2022.

Fama, E.F., French, K.R. 2015. A Five-Factor Asset Pricing Model. *Journal of Financial Economics* 116: 1–22. https://doi.org/10.1016/j.jfineco.2014.10.010. Accessed 2 February 2021.

Gordon, M.J. 1959. Dividends, Earnings, and Stock Prices. *The Review of Economics and Statistics* 41: 99–105. https://doi.org/10.2307/1927792. Accessed 2 February 2021.

Gordon, J.R., Gordon, M.J. 1997. The Finite Horizon Expected Return Model. *Financial Analysts Journal* 53: 52–61. https://doi.org/10.2469/faj.v53.n3.2084. Accessed 2 February 2021.

Gordon, M.J., Shapiro, E. 1956. Capital Equipment Analysis: The Required Rate of Profit. *Management Science* 3: 102–110. https://www.jstor.org/stable/2627177. Accessed 2 February 2021.

Graham, J., Harvey, C. 2002. How CFOs Make Capital Budgeting and Capital Structure Decisions. *Journal of Applied Corporate Finance* 15: 8–23. https://doi.org/10.1111/j.1745-6622.2002.tb00337.x. Accessed 2 February 2021.

Harvey, R., Liu, Y., Zhu, H. 2016. ... and the Cross-Section of Expected Returns. *The Review of Financial Studies* 29: 5–68. https://doi.org/10.1093/rfs/hhv059. Accessed 2 February 2021.

Isaac, D., O'Leary, J., 2013. *Property Valuation Techniques*. 3rd Ed. London: Palgrave Macmillan.

Koller, T., Goedhart, M. Wessels, D., McKinsey and Company. 2015. *Valuation: Measuring and Managing the Value of Companies*. 6th Ed. New Jersey: Wiley.

Koller, T., Dobbs, R., Huyett, B., McKinsey and Company. 2011. *Value: The Four Cornerstones of Corporate Finance*. 6th Ed. New Jersey: Wiley.

Lintner, J. 1965. The Valuation of Risk Assets and the Selection of Risky Investments in Stock Portfolios and Capital Budgets. *The Review of Economics and Statistics* 47: 13–37. https://doi.org/10.2307/1924119. Accessed 2 February 2023.

Markowitz, H. 1952. Portfolio Selection. *The Journal of Finance* 7: 77–91. https://doi.org/10.2307/2975974. Accessed 2 February 2023.

Merton, R. 1973. An Intertemporal Capital Asset Pricing Model. *Econometrica* 41: 867–887. https://doi.org/10.2307/1913811. Accessed 2 February 2023.

Modigliani, F., Miller, M.H. 1958. The Cost of Capital, Corporation Finance and the Theory of Investment. *The American Economic Review* 48: 261–297. https://www.jstor.org/stable/1809766. Accessed 2 February 2023.

Modigliani, F., Miller, M.H. 1963. Corporate Income Taxes and the Cost of Capital: A Correction. *The American Economic Review* 53: 433–443. https://www.jstor.org/stable/1809167. Accessed 2 February 2023.

Nobel Prize. 1997. For a new method to determine the value of derivatives. The Nobel Prize. Press Release. https://www.nobelprize.org/prizes/economic-sciences/1997/press-release/. Accessed 02 February 2022.

Papazian, A. 2022. *The Space Value of Money: Rethinking Finance Beyond Risk and Time*. New York: Palgrave Macmillan. https://doi.org/10.1057/978-1-137-59489-1.

Papazian, A. 2023. *Hardwiring Sustainability into Financial Mathematics: Implications for Money Mechanics*. New York: Palgrave Macmillan. https://doi.org/10.1007/978-3-031-45689-3.

Pike, R., Neale, B., Akbar, S., Linslley, P. 2018. *Corporate Finance and Investment*. 9th Ed. London: Pearson.

Roll, R., Ross, S.A. 1980. An Empirical Investigation of the Arbitrage Pricing Theory. *The Journal of Finance* 35: 1073–1103.

Rosenbaum, J. Pearl, J. 2013. *Investment Banking*. New Jersey: Wiley.

Ross, S.A. 1976. The Arbitrage Theory of Capital Asset Pricing. *Journal of Economic Theory* 13: 341–360. https://doi.org/10.1016/0022-0531(76)90046-6. Accessed 2 February 2023.

Sharpe, W.F. 1964. Capital Asset Prices: A Theory of Market Equilibrium Under Conditions of Risk. *Journal of Finance* 19: 425–42. https://doi.org/10.1111/j.1540-6261.1964.tb02865.x. Accessed 2 February 2023.

Watson, D., Head, A. 2016. *Corporate Finance: Principles and Practice*. 7th ed. London: Pearson.

Williams, J.B. 1938. *The Theory of Investment Value*. Cambridge: Harvard University Press.

Yescombe, E.R. 2014. *Principles of Project Finance*. 2nd Ed. Oxford: Academic Press.

7

Spaceless Debt-Based Monetary Architecture

We went to the moon as technicians; we returned as humanitarians.
Edgar Mitchell, Apollo 14 Astronaut, 1971

Going to space, every one of us went as a patriot of our own country. But we came back as patriots of our Earth.
Fyodor Yurchikhin Grammatikopoulos, STS-112 Atlantis Astronaut, 2002

Having revealed a space blind financial value framework and mathematics, our discussion must now address our monetary architecture. This chapter is about how money is created. It is not an academic and historical review of money and money creation, but a factual description based mainly on Bank of England and US Federal Reserve publications. I have adopted this approach intentionally. The purpose is not to debate the many theoretical functions of money or the role of banks as financial intermediaries. My aim here is to demonstrate that money is created through debt instruments.

After discussing the debt-based nature of money, I introduce three major systemic limitations caused by our debt-based monetary architecture. However, before expanding on these limitations, which I do in the next three chapters, I briefly discuss cryptocurrencies. This is to establish the fact that they are equally spaceless and unsustainable.

© The Author(s), under exclusive license to Springer Nature
Switzerland AG 2024
A. V. Papazian, *Financing the Race to Space*,
https://doi.org/10.1007/978-3-031-73102-0_7

7.1   Debt-Based Money

Our current monetary system is debt-based. Money, in all its forms, is created through debt transactions and instruments. McLeay et al. (2014a) describe our debt-based monetary system in an article published in the Bank of England Quarterly Bulletin as follows:

> There are three main types of money: currency, bank deposits and central bank reserves. Each represents an IOU from one sector of the economy to another. Most money in the modern economy is in the form of bank deposits, which are created by commercial banks themselves. (McLeay et al., 2014a, 4)

Figure 7.1 depicts the debt/loan-based money creation process between central banks, commercial banks, and consumers. It is adapted verbatim from McLeay et al. (2014b). The figure reveals the basic fact that the three forms of money, currency, deposits, and central bank reserves are in fact created through debt instruments and transactions, or IOUs (I owe you).

> '[B]road money' circulating in the economy... can be thought of as the money that consumers have available for transactions, and comprises: currency (banknotes and coin)—an IOU from the central bank, mostly to consumers in the economy; and bank deposits—an IOU from commercial banks to consumers....
> '[B]ase money' or 'central bank money', comprises IOUs from the central bank: this includes currency (an IOU to consumers) but also central bank reserves, which are IOUs from the central bank to commercial banks (McLeay et al. 2014a, 7)

In a subsequent publication, again in the Bank of England Quarterly Bulletin, McLeay et al. (2014b) describe how commercial banks create money through loans.

> In the modern economy, most money takes the form of bank deposits. But how those bank deposits are created is often misunderstood: the principal way is through commercial banks making loans. Whenever a bank makes a loan, it simultaneously creates a matching deposit in the borrower's bank account, thereby creating new money. (McLeay et al. 2014b, 1)

At the Central Bank level, looking at the Balance Sheet of the Federal Reserve, Table 7.1, we observe that Federal Reserve Notes (US Dollars) and Central Bank Reserves (deposits held by depository institutions) are liabilities backed by assets that are all debt instruments.

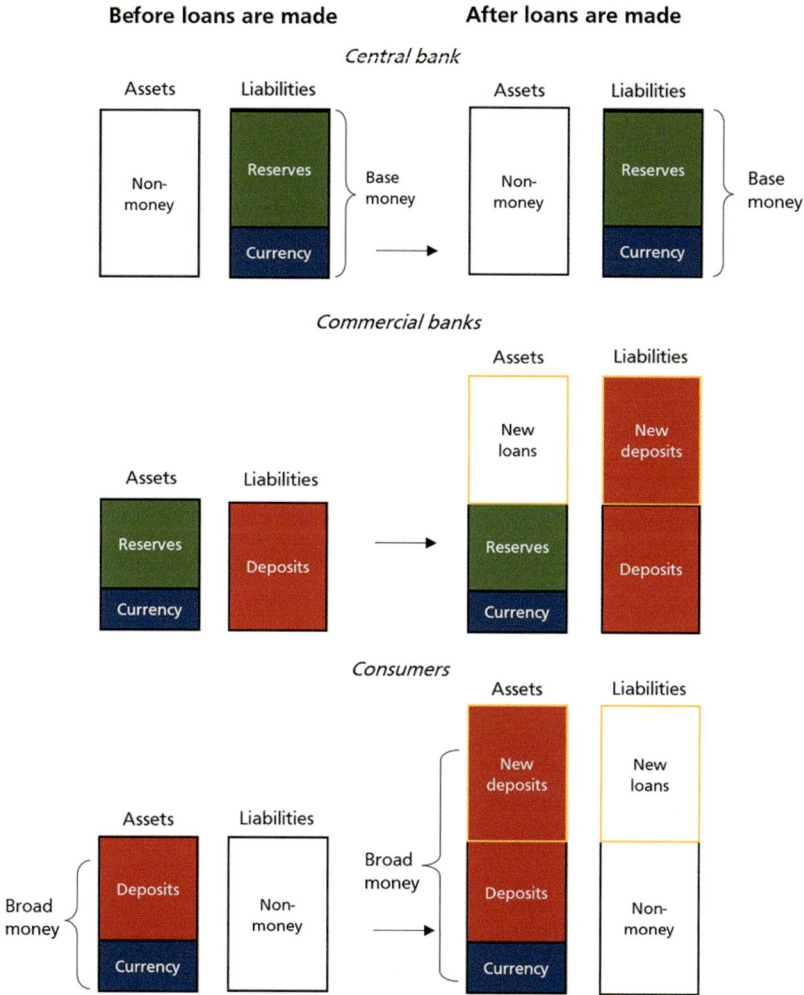

Fig. 7.1 Debt-based money creation process (*Source* Adapted from McLeay et al. [2014b])

One of the most commonly discussed instance of money creation by central banks is called Quantitative Easing (QE) or Credit Easing (CE). QE was used in the 2008 financial crisis and the 2020 corona virus pandemic to support markets and economies in the US, UK, Europe, and elsewhere. QE refers to the strategy of creating money or liquidity through the purchase of investment grade assets from primary dealers. This method of new money injection, or liquidity injection, increases the amount of money in the banking system through increased central bank reserves. This does not affect the number of printed banknotes in circulation as such, and does not at first

Table 7.1 Assets, liabilities, and capital of the Federal Reserve system ($billions)

	March 27, 2024	September 27, 2023
Total assets	**7,485**	**8,002**
Securities held outright	7,009	7,440
U.S. Treasury securities	4,618	4,958
Federal agency debt securities	2	2
Agency mortgage-backed securities	2,388	2,480
Repurchase agreements	0	0
Foreign official	0	0
Other	0	0
Loans	142	198
Discount window	6	3
Bank Term Funding Program	133	108
Paycheck Protection Program Liquidity Facility	3	5
Other credit extensions	0	82
Net portfolio holdings of Main Street Facilities LLC	15	19
Net portfolio holdings of Municipal Liquidity Facility LLC	0	6
Net portfolio holdings of Term Asset-Backed Securities Loan Facility II LLC	0	1
Central bank liquidity swaps	0	0
Other assets	319	337
Total liabilities	**7,442**	**7,959**
Federal Reserve notes	2,293	2,273
Deposits held by depository institutions other than term deposits	3,472	3,169
Reverse repurchase agreements	873	1,755
Foreign official and international accounts	354	312
Others	518	1,443
U.S. Treasury, General Account	772	672
Treasury contributions to credit facilities	7	13
Other liabilities	24	77
Total capital	**43**	**43**

Source Fed (2024a)

affect commercial bank deposits, although it may at later stages when/if new central bank reserves translate into new loans to consumers.

Chart 7.1 depicts the Asset side of the Federal Reserve balance sheet. The chart clearly shows the jump in assets thanks to the purchase of securities used as part of the QE strategy. The Bank of England describes the QE process simply and eloquently:

The money we used to buy bonds when we were doing QE did not come from government taxation or borrowing. Instead, like other central banks, we can create money digitally in the form of 'central bank reserves'. We use these reserves to buy bonds. Bonds are essentially IOUs issued by the government and businesses as a means of borrowing money. (Bank of England 2024)

Chart 7.2 depicts selected liabilities of the FED reflecting the changes in the Asset side shown in Chart 7.1 We can observe the commensurate jump in central bank reserves, deposits of depository institutions, and the treasury balance. Chart 7.3 depicts the combined picture identifying key stages of QE from 2007 to date. QT refers to Quantitative Tightening, which is the exact opposite process of selling securities and tightening the money supply. Table 7.2 provides a list of the Primary Dealers of the Federal Reserve Bank of New York (NYFED 2024). These primary dealers are the institutions, trading counterparties, used by the New York Federal Reserve when implementing its monetary policy. In other words, they are the intermediaries through which the debt instruments are purchased with new liquidity

On the 23rd of March 2020, the US Federal Reserve announced a targeted initiative to support the US corporate bond market, as part of its response to the pandemic and troubled markets. It created the Primary Market Corporate Credit Facility (PMCCF) and the Secondary Market Corporate Credit Facility (SMCCF). These were unprecedented direct interventions in the corporate bond market (Bernanke 2009, FED 2020).

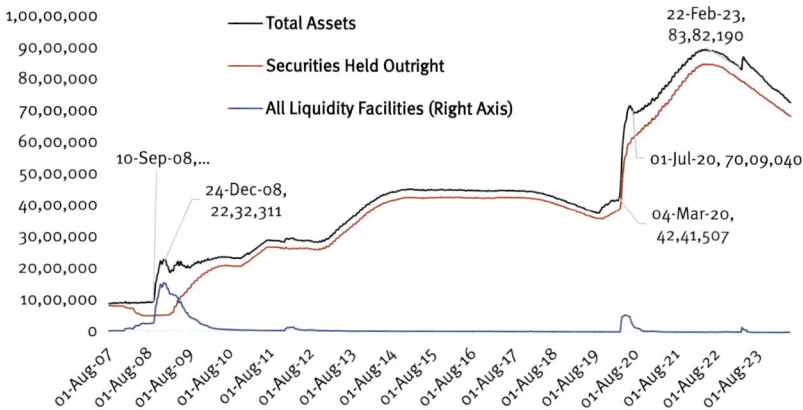

Chart 7.1 Federal Reserve Balance Sheet, total assets in $millions (*Source* FED [2024b])

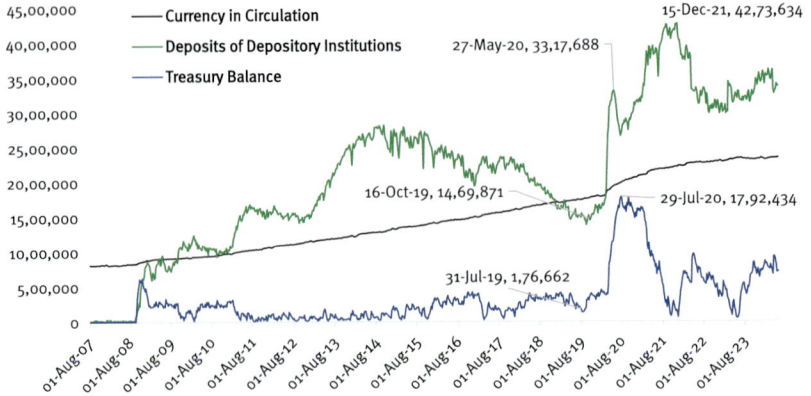

Chart 7.2 Federal Reserve Balance Sheet, selected liabilities in $millions (*Source* FED [2024c])

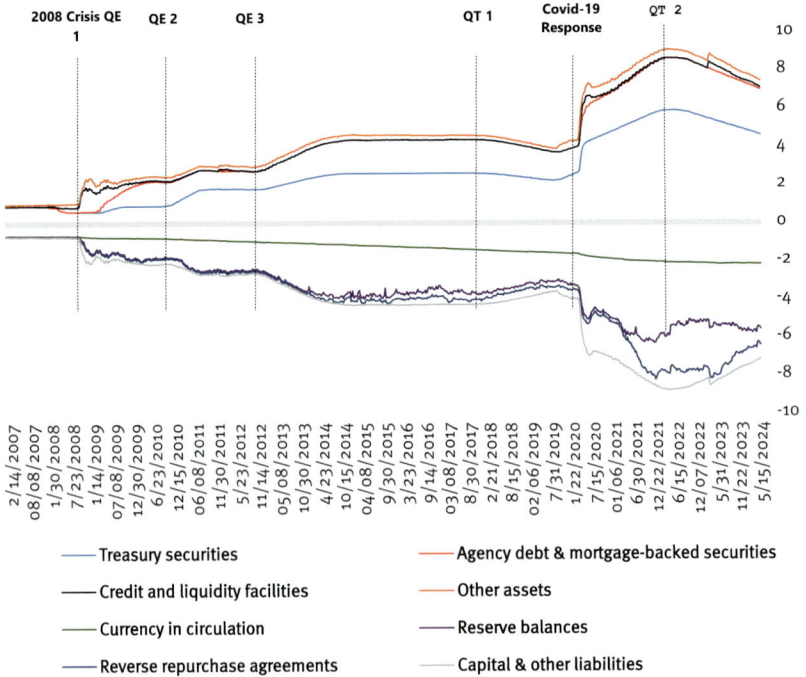

Chart 7.3 Federal Reserve assets & liabilities, weekly in trillions of dollars, 2007–2024 (*Source* FED [2024d])

Table 7.2 New York Federal Reserve primary dealers list

ASL Capital Markets Inc	Jefferies LLC
Bank of Montreal, Chicago Branch	J.P. Morgan Securities LLC
Bank of Nova Scotia, New York Agency	Mizuho Securities USA LLC
BNP Paribas Securities Corp	Morgan Stanley & Co. LLC
Barclays Capital Inc.	NatWest Markets Securities Inc
BofA Securities, Inc.	Nomura Securities International, Inc
Cantor Fitzgerald & Co	RBC Capital Markets, LLC
Citigroup Global Markets Inc.	Santander US Capital Markets LLC
Daiwa Capital Markets America Inc.	Société Generale, New York Branch
Deutsche Bank Securities Inc.	TD Securities (USA) LLC
Goldman Sachs & Co. LLC	UBS Securities LLC
HSBC Securities (USA) Inc.	Wells Fargo Securities, LLC

Primary dealers are trading counterparties of the New York Fed in its implementation of monetary policy. They are also expected to make markets for the New York Fed on behalf of its official accountholders as needed, and to bid on a pro-rata basis in all Treasury auctions at reasonably competitive prices
Source NYFED (2024)

The PMCCF provided companies access to credit so that they were better able to maintain business operations and capacity during the period of dislocations related to the pandemic. This facility was open to investment grade companies, as well as certain companies that were investment grade as of March 22, 2020. The Federal Reserve established a special purpose vehicle (SPV) through which the PMCCF was able to make loans and purchase bonds. (FED 2021)

As you can see from the above quote, the facility was open to investment grade companies or bonds. This is very important to note because it also reveals that the focus is on creditworthiness of the issuing corporates, not on their space impact. The absence of space parameters, as discussed in Chapters 5 and 6, remains relevant to all the instruments used for injecting new money into the economies. To better understand what investment grade bonds mean, in Table 7.3, you can find the long-term and short-term rating scale of Moody's (2023), where you can see the ratings and their definitions. The entire exercise is focused on ranking issuers based on their level of credit risk.

Looking at the Bank of England QE programme from 2009 to 2021, we observe the same experience, where government bonds (gilts) and corproate bonds were purchased to enable the injection of liquidity, or the creation of new money in the banking system. Once again, the focus is not on what the bond does or what impact it has, but on the reliability of its cash flows. Indeed, the Bank of England and the US Federal Reserve published their respective list of eligible bonds (debt instruments) that could be used in their

Table 7.3 Moody's ratings scale and definitions

	Long-term ratings	Long-term rating definitions	Short-term ratings	Short-term rating definitions
Investment grade	Aaa	Obligations rated Aaa are judged to be of the highest quality, subject to the lowest level of credit risk	P-1	Issuers (or supporting institutions) rated Prime-1 (P-1) have a superior ability to repay short-term debt obligations
	Aa1	Obligations rated Aa are judged to be of high quality and are subject to very low credit risk	P-1	
	Aa2		P-1	
	Aa3		P-1	
	A1	Obligations rated A are considered upper-medium grade and are subject to low credit risk	P-1	
	A2		P-2, P-1	Issuers (or supporting institutions) rated Prime-2 (P-2) have a strong ability to repay short-term debt obligations
	A3		P-2, P-1	

	Long-term ratings	Long-term rating definitions	Short-term ratings	Short-term rating definitions
	Baa1 Baa2 Baa3	Obligations rated Baa are judged to be medium grade and subject to moderate credit risk and as such may possess certain speculative characteristics	P-2 P-3, P-2 P-3	Issuers (or supporting institutions) rated Prime-3 (P-3) have an acceptable ability to repay short-term debt obligations
Speculative grade	Ba1 Ba2 Ba3	Obligations rated Ba are judged to be speculative and are subject to substantial credit risk	NP (Not Prime)	Issuers (or supporting institutions) rated Not Prime (NP) do not fall within any of the Prime rating categories
	B1 B2 B3	Obligations rated B are considered speculative and are subject to high credit risk		

(continued)

Table 7.3 (continued)

Long-term ratings	Long-term rating definitions	Short-term ratings	Short-term rating definitions
Caa1 Caa2 Caa3	Obligations rated Caa are judged to be of poor standing and are subject to very high credit risk. Elements are subject to very high credit risk		
Ca	Obligations rated Ca are highly speculative and are likely in, or very near, default, with some prospect of recovery in principal and interest		
C	Obligations rated C are the lowest-rated class of bonds and are typically in default, with little prospect for recovery of principal and interest		

Note Moody's appends numerical 1, 2, and 3 to each generic rating classification from Aa to Caa. The modifier 1 indicates that the obligation ranks in the higher end of its generic rating category; the modifier 2 indicates a mid-range ranking; and the modifier 3 indicates a ranking in the lower end of the generic rating category

Source Moody's (2023)

Quantitative Easing programmes (Bank of England 2021b; FEDNY 2020). Both lists echo this emphasis (Fig. 7.2).[1]

Naturally, credit ratings are also used by commercial banks. Individual and business ratings are key factors that determine access to credit. One additional element that often comes into play when commercial banks lend to individuals, households, and businesses is the provision of an asset as collateral. This is widely reflected in the application of the Loan-To-Value (LTV) ratio and its use by commercial banks (Lloyds 2024; HSBC 2024). LTV ratios define the percentage of the value of the collateral asset that banks are willing to lend to a specific borrower. It usually depends on the 5 Cs of credit or creditworthiness, i.e., character, capacity, capital, collateral, and conditions (Papazian 2022, 222). Once again, the focus is on the reliability of expected cash flows, and not on their space impact.

The above discussion aimed to demonstrate that money, in all its forms, i.e., currency, deposits, and central bank reserves, is created and injected through debt instruments, i.e., credit, loans, and bonds. Table 7.4 provides a sample summary list of debt instruments and transactions used for creating money. These are not exhaustive and do not necessarily apply to all.

7.1.1 Three Systemic Limitations

The debt-based logic of money creation imposes unique constraints and systemic limitations on our economies and plays a critical role in shaping the two challenges we started with: our inability to invest and build sustainably on Earth, and our inability to invest and expand in outer space. I identify and discuss three systemic challenges caused by our spaceless debt-based monetary architecture where money creation is built on debt instruments designed and valued in a risktime universe without space, without outer space. They are:

(a) Calendar time: a muzzle in space
(b) Monetary gravity: a leash in space
(c) Monetary hunger: a whip in space

---

[1] Interestingly, during a Bank of England Future Forum on Money, in 2018, to which I participated in person, I was witness to this very discussion regarding the selection criteria of bonds used for monetisation, for money injection. I raised this issue personally with the then Governor Mark Carney (Bank of England 2018, 1:23:00), sharing the high-level insights that I discuss in this and previous books (Papazian 2023, 2022).

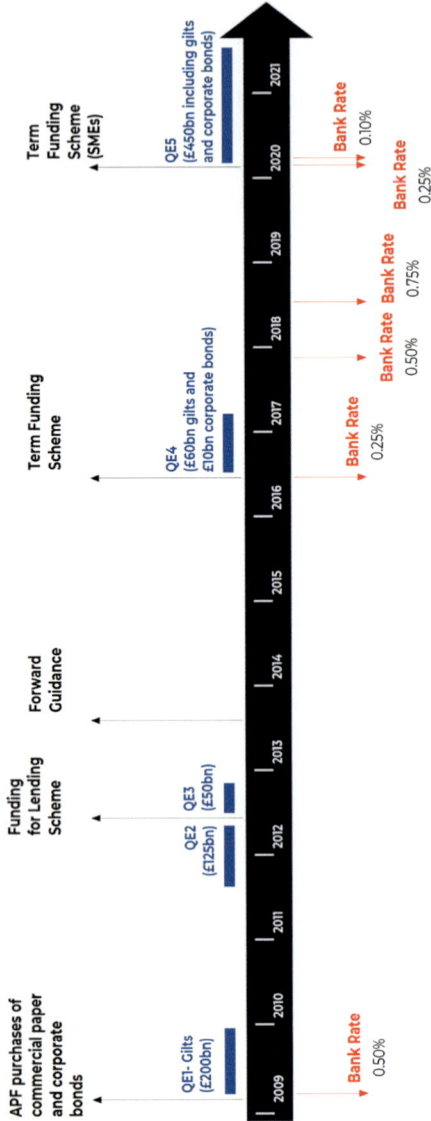

Fig. 7.2 Bank of England QE programmes and selected policy interventions since 2009 (*Source* Adapted from Bank of England [2021a])

Table 7.4 Sample debt instruments, portfolios, and transactions

	Commercial banks	Central banks
Instruments	Consumer Credit Business Credit Residential Mortgages Commercial Mortgages	Discount Loans TLTRO Loans Subsidiary Loans
Portfolios	Loan Portfolios Mortgage Portfolios	Government Bond Portfolios[2] Corporate Bond Portfolios MBS and CDO Portfolios Commercial Paper

Transactional engagements	Loan Approvals
	Mortgage Approvals
		Currency Issuance
		Reserve Issuance
		  – Quantitative Easing
		  – Credit Easing

Source Papazian (2023)

I explore these challenges in detail in the following chapters. In the next section, I discuss cryptocurrencies, albeit very briefly, in order to establish the fact that they are not a viable and sustainable alternative that can replace our current monetary architecture. Moreover, space and space impact are still unaccounted for in this new form of 'money.'

7.2   Cryptocurrencies

Despite their popularity in some circles, cryptocurrencies[3] are not really an improvement on debt-based money. As the reader will discover in the following chapters, the shortcomings of debt-based money are severe and have evolutionary implications. But, as the below discussion will demonstrate, cryptocurrencies are not actual currencies, they have their own set of issues, and their logic of creation does not address the absence of space, and thus outer space. In other words, while the cryptocurrency fascination may reflect the need and push for radical reform in our monetary system, Bitcoin and other cryptoassets are not necessarily and actually in themselves a reliable and sustainable alternative to debt-based fiat money. As I discuss at length in

---

[2] Note for clarification that while commercial bank loans create new deposits directly (new money), government and corporate bonds issued and sold in capital markets do not, they are purchased with existing money. However, bonds are used by central banks to create new central bank reserves (new money) through their monetary policy injections, or QE strategy.

[3] Cryptocurrencies should not be confused with Central Bank Digital Currencies (CBDC) which are now being considered by the Bank of England and other central banks (Bank of England 2023a).

later chapters, the opportunity for reform is real and it is about a transformed financial value framework, mathematics, and monetary architecture.

As of the 4th of June 2024, there were 10,035 cryptocurrencies listed on coinmarketcap.com (CoinMarketCap 2024). This number is as volatile as the prices of the listed cryptocoins. On the top of the list is Bitcoin, and as I have done so previously, I use it as an example in this discussion.

The first issue to address is the fact that cryptocurrencies are not really currencies. The Bank of England refers to them as cryptoassets:

> Put it this way, you wouldn't use cryptocurrency to pay for your food shop. In the UK, no major high street shop accepts cryptocurrency as payment. It's generally slower and more expensive to pay with cryptocurrency than a recognised currency like sterling. Development is underway to make cryptocurrency easier to use, but for now it isn't very 'money-like'. This is why central banks now refer to them as "cryptoassets" instead of "cryptocurrencies". Today cryptocurrencies are generally held as investments by people who expect their value to rise. (Bank of England, 2020)

Another key issue of cryptoassets concerns the essential dynamic of their price volatility. Prasad (2021) addresses this referring to the 'greater fool' theory. He writes: "The valuations of meme currencies seem to be based entirely on the "greater fool" theory—all you need to do to profit from your investment is to find an even greater fool willing to pay a higher price than you paid for the digital coins." A competing interpretation of their use and attractiveness is explained by The Economist: it is their lack of transparency and their usefulness as 'dark money' (*Economist* 2022).

On top of the above discussed challenges, i.e., volatility, lack of transparency, and absence of any intrinsic value, Bitcoin, and other cryptoassets are limited in number and also have a significant electronic waste footprint due to the mining process involved in using and generating the coins. According to a recent study done by De Vries and Stoll (2021), the e-waste of Bitcoin is estimated at 30.7 metric kilotons per year as of May 2021.

Similarly and in parallel, the Cambridge Bitcoin Energy Consumption Index (CBECI) estimates that the yearly average annualised electricity consumption of the Bitcoin Network is around 135.42 TWh per year, higher than the yearly consumption of Ukraine at 134.3 TWh per year (CCAF 2024).[4]

---

[4] For a relative understanding of the numbers, note that the two highest consuming countries, China and USA, use respectively 7,805.65 TWh and 3,979.27 TWh per year (CCAF 2024).

While all of the above-identified issues are relevant in determining the true nature and relevance of cryptoassets, the most defining challenge is their very logic of creation. Bitcoins are created after a mining process through powerful computers performing specific mining operations which consist in solving complex mathematical puzzles.

Anybody can become a Bitcoin miner by running software with specialised hardware. Mining software listens for transactions broadcast through the peer-to-peer network and performs appropriate tasks to process and confirm these transactions. Bitcoin miners perform this work because they can earn transaction fees paid by users for faster transaction processing, and newly created bitcoins issued into existence according to a fixed formula.

For new transactions to be confirmed, they need to be included in a block along with a mathematical proof of work. **Such proofs are very hard to generate because there is no way to create them other than by trying billions of calculations per second**. This requires miners to perform these calculations before their blocks are accepted by the network and before they are rewarded. As more people start to mine, the difficulty of finding valid blocks is automatically increased by the network to ensure that the average time to find a block remains equal to 10 minutes. As a result, mining is a very competitive business where no individual miner can control what is included in the block chain. (Bitcoin 2023a, b)[5]

In other words, while fiat money, i.e., debt-based money, is created via debt instruments and transactions, Bitcoins are created/awarded through a process that involves and is dependent on '*trying billions of calculations per second.*' Bitcoin replaces the debt logic of fiat money with mathematical guesswork. Moreover, cryptocurrencies or cryptoassets do not address the absence of space, and they do not include space impact as an integral part of the money creation process. They are, thus, equally detached from our physical context and our impact on it.

---

[5] Emphasis added.

7.3   Conclusion

This chapter demonstrated that money in all its forms, i.e., currency, deposits, and central bank reserves, is created through debt instruments. I argued that replacing the 'unwanted servitude' of debt-based money with the 'mathematical randomness' of Bitcoin is not really an improvement. Moreover, sustainability and space are still unaccounted for.

The next part and three chapters dive deeper into the three structural implications of debt-based money and the limitations they impose on us in space: (1) Calendar Time, (2) Monetary Gravity, and (3) Monetary Hunger.

References

Bank of England. 2018. Bank of England Future Forum on Money. Bank of England. https://www.youtube.com/watch?v=--8t85ipwvo&t=4973s. At 1:23:00 min. Accessed 12 May 2024.

Bank of England. 2020. What Are Cryptoassets (Cryptocurrencies)? Bank of England, England. https://www.bankofengland.co.uk/knowledgebank/what-are-cryptocurrencies. Accessed 2 February 2024.

Bank of England. 2021a. IEO Evaluation of the Bank of England's Approach to Quantitative Easing. Bank of England. https://www.bankofengland.co.uk/independent-evaluation-office/ieo-report-january-2021/ieo-evaluation-of-the-bank-of-englands-approach-to-quantitative-easing. Accessed 12 March 2024.

Bank of England. 2021b. Bank of England Corporate Bond Purchase Scheme: Eligible Bonds List. Bank of England. https://www.bankofengland.co.uk/-/media/boe/files/markets/corporate-bond-purchases/bonds-eligible-for-the-corporate-bond-purchase-scheme.xlsx. Accessed 5 June 2024.

Bank of England. 2021c. Bank of England Annual Report and Accounts 20/21. Bank of England. https://www.bankofengland.co.uk/-/media/boe/files/annual-report/2021/boe-2021.pdf#page=97. Accessed 2 February 2024.

Bank of England. 2023a. What Is CBDC? https://www.bankofengland.co.uk/explainers/what-is-a-central-bank-digital-currency. Accessed 1 March 2023.

Bank of England. 2023b. Bank of England Annual Report and Accounts 22/23. Bank of England. https://www.bankofengland.co.uk/-/media/boe/files/annual-report/2023/boe-2023.pdf#page=201. Accessed 1 February 2024.

Bank of England. 2024. Quantitative Easing. The Bank of England. https://www.bankofengland.co.uk/monetary-policy/quantitative-easing. Accessed 12 June 2024.

Bernanke, S.B. 2009. The Crisis and the Policy Response. Federal Reserve Board of Directors. Speech at London School of Economics. https://www.federalreserve.gov/newsevents/speech/bernanke20090113a.htm. Accessed 2 February 2023.

Bitcoin. 2023a. How Are Bitcoins Created? Bitcoin.org. https://bitcoin.org/en/faq# how-are-bitcoins-created. Accessed 12 December 2023.

Bitcoin. 2023b. How Does Bitcoin Mining Work? Bitcoin.org. https://bitcoin.org/ en/faq#how-does-bitcoin-mining-work. Accessed 12 December 2021.

CCAF. 2024. *Cambridge Bitcoin Electricity Consumption Index.* Cambridge: Cambridge Centre for Alternative Finance. https://ccaf.io/cbnsi/cbeci/compar isons. Accessed 28 June 2024.

CoinMarketCap. 2024. *All Cryptocurrencies.* CoinMarketCap. https://coinmarke tcap.com/. Accessed 4 June 2024.

De Vries, A., Stoll, C. 2021. Bitcoin's Growing E-waste Problem. Resources Conservation and Recycling. https://doi.org/10.1016/j.resconrec.2021.105901. Accessed 2 February 2022.

Economist. 2022. The Charm of Cryptocurrencies for White Supremacists. *The Economist.* https://www.economist.com/united-states/2022/02/05/the-charm-of-cryptocurrencies-for-white-supremacists. Accessed 22 February 2022.

FED. 2020. Federal Reserve Announces Extensive New Measures to Support the Economy. https://www.federalreserve.gov/newsevents/pressreleases/moneta ry20200323b.htm. Accessed 12 February 2024.

FED. 2021. Primary Market Corporate Credit Facility. US Federal Reserve. https:// www.federalreserve.gov/monetarypolicy/pmccf.htm. Accessed 12 April 2024.

FED. 2024a. Federal Reserve Balance Sheet Developments. US Federal Reserve. https://www.federalreserve.gov/publications/files/balance_sheet_develop ments_report_202405.pdf. Accessed 12 June 2024.

FED. 2024b. Balance Sheet Trends—Total and Selected Assets. US Federal Reserve. https://www.federalreserve.gov/monetarypolicy/bst_recenttrends_access ible.htm. Accessed 31 May 2024.

FED. 2024c. Balance Sheet Trends—Selected Liabilities. US Federal Reserve. https://www.federalreserve.gov/monetarypolicy/bst_recenttrends_accessible.htm. Accessed 31 May 2024.

FED. 2024d. Federal Reserve Assets and Liabilities. US Federal Reserve Board. https://www.federalreserve.gov/newsevents/speech/bowman20240528a1. htm. Accessed 31 May 2024.

FEDNY. 2020. Composition of the SMCCF Broad Market Index. Federal Reserve Bank of New York. https://www.newyorkfed.org/markets/secondary-market-cor porate-credit-facility/secondary-market-corporate-credit-facility-broad-market-index. Accessed 12 April 2024.

HSBC. 2024. 95% Mortgages: Helping First Time Buyers Get on the Property Ladder. HSBC Holdings PLC. https://www.hsbc.co.uk/mortgages/95-percent/. Accessed 4 June 2024.

Lloyds. 2024. What Is Loan to Value Ratio? Lloyd's Banking Group. https://www. lloydsbank.com/mortgages/help-and-guidance/moving-home/what-is-loan-to-value-ratio.html. Accessed 12 May 2024.

McLeay, M., Radia, A., Thomas, R. 2014a. Money in the Modern Economy: An Introduction. Bank of England. Quarterly Bulletin. https://www.bankofeng

land.co.uk/-/media/boe/files/quarterly-bulletin/2014/money-in-the-modern-eco nomy-an-introduction.pdf. Accessed 6 June 2023.

McLeay, M., Radia, A., Thomas, R. 2014b. Money Creation in the Modern Economy. Bank of England. Quarterly Bulletin. https://www.bankofengland.co. uk/-/media/boe/files/quarterly-bulletin/2014/money-creation-in-the-modern-eco nomy. Accessed 6 June 2020

Moody's. 2023. Rating Symbols and Definitions. Moody's Investor Service. https:// ratings.moodys.com/rmc-documents/53954. Accessed 12 April 2024.

NYFED. 2024. Primary Dealers. Federal Reserve Bank of New York. https://www. newyorkfed.org/markets/primarydealers. Accessed 12 May 2024.

Papazian, A. 2022. *The Space Value of Money: Rethinking Finance Beyond Risk and Time*. New York: Palgrave Macmillan. https://doi.org/10.1057/978-1-137-594 89-1.

Papazian, A. 2023. *Hardwiring Sustainability into Financial Mathematics: Implica- tions for Money Mechanics*. New York: Palgrave Macmillan. https://doi.org/10. 1007/978-3-031-45689-3.

Prasad, E. 2021. Five Myths About Cryptocurrency. *The Washington Post*. https:// www.washingtonpost.com/outlook/five-myths/cryptocurrency-yths-bitcoin-dog ecoin-musk/2021/05/20/1f3f6c28-b8ad-11eb-96b9-e949d5397de9_story.html. Accessed 2 February 2022.

Part III

Muzzle, Leash, and Whip in Space

This part of the book explores the three systemic shortcomings of our debt-based monetary architecture identified in the previous chapter: calendar time, monetary gravity, and monetary hunger. These structural challenges can further explain our inability to invest and build sustainably on Earth and our inability to invest and expand in outer space.

8

Calendar Time: A Muzzle in Space

I don't know what you could say about a day in which you have seen four
beautiful sunsets.
John Glenn, Mercury Friendship 7 Astronaut, 1962

The more people that go into space, the better the future of humanity will be.
Marc Garneau, STS-41-G Challenger Astronaut, 1984

Commercial bank loans, central bank loans, and bond purchases are the
main tools of money creation in our current monetary architecture. Our
spaceless financial value framework and financial mathematics define the
analytical principles and tools we use to value the debt instruments upon
which our spaceless monetary architecture is built. Using cashflow or bond
valuation equations, debts are assessed within and through risk and time
parameters alone. At the heart of the instruments we use for money creation,
and the framework we apply to assess their value, is calendar time.

This chapter is dedicated to demonstrating that this innocent assumption,
using calendar time as a fundamental pillar of our monetary architecture,
acts as a muzzle on our ability to invest and expand in outer space. This is so
because a monetary architecture built around calendar time limits our ability
to create and invest the necessary monetary resources *timelessly*.

© The Author(s), under exclusive license to Springer Nature
Switzerland AG 2024
A. V. Papazian, *Financing the Race to Space*,
https://doi.org/10.1007/978-3-031-73102-0_8

8.1   Debt Instruments and Valuation

This discussion is contextualised using bonds as an example, but the key proposition applies to all kinds of loans and debts. Looking at bonds specifically (Fig. 8.1 and Eq. 7.1), we can identify a number of key elements that define the instrument as well as the equations we use to assess its value.

The main elements of a bond include the Par Value (P) or Principal Amount, which is the amount to be returned to the bondholder at maturity, the Coupon Rate, which determines the periodic payments (C_t) to be paid to the bondholder as a percentage of the par value, the date to Maturity (n), which is the date when the par value is repaid to the bondholder, and the discount rate or Yield to Maturity (r), which is the return that an investor can achieve if they reinvested all coupon payments at the same rate as the bond's return and held the bond to maturity. The discount rate, also known as redemption yield, is used to measure the present value of the cash flows expected from the bond. In other words, the future payments are discounted to the present to determine their risk and time value for the investor in the present.

As we can observe, the entire formula is built around time (t) which starts in the present and moves to the end period (n), and Space, as already discussed, is absent. The impact of the bond is omitted from the equation that assesses its value. Thus, and as the reader knows by experience, one of the most central pillars of all debt instruments, including bonds and loans, secured or unsecured, with collateral or without, is the maturity date and the schedule of repayment defined by calendar time. At the heart of a debt-based monetary architecture and the instruments used to create money, is a social

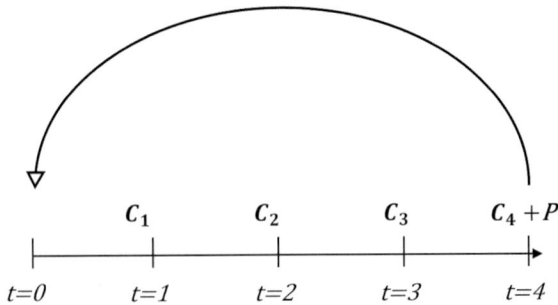

Fig. 8.1 Bond payout cash flow timeline (*Source* Author)

construct that we take for granted, i.e., calendar time.[1]

$$\text{Bond Value} = \sum_{t=1}^{n} \frac{C_t}{(1+r)^t} + \frac{P}{(1+r)^n} \tag{8.1}$$

$$n = \text{Time to Maturity}$$

$$t = \text{Moving time}$$

$$r = \text{Discount Rate or Yield to Maturity}$$

$$P = \text{Par Value of Bond}$$

$$C_t = \text{Coupon Payments}$$

Government and corporate debts, business and personal loans, and all credit instruments have a calendar time-based maturity and repayment schedule. Creating money through instruments that are linked to calendar time chains the entire money supply to the rotation and revolution of Earth around the sun. As I elaborate in the next section, this is so because the two most important units of our calendar, a day and a year, are based respectively on the rotation of Earth on itself and the revolution of Earth around the sun.

This use of calendar time as a fundamental pillar of our monetary architecture acts as a muzzle on our ability to invest and expand in outer space because it limits our ability to create and invest the necessary amounts of money *outside this social construct.* Our calendar time dependent monetary architecture chains us to a fixed and limited measurement of Earth's movements in outer space, specifically its rotation and revolution around the sun, undermining our ability to create the resources we need to explore and settle in worlds that are at great distances from our own, and have very different calendars and orbits of their own. In the vast landscape of space, calendar time linked debt-based money undermines our ability to explore what is relatively limitless, i.e., space.

---

[1] The list of all the corporate bonds eligible for the Bank of England QE program also lists their maturity date (Bank of England 2021).

8.2   Calendar Time

> Debt, which involves time obligations in terms of scheduled interest and principal repayments, chains everybody involved to calendar time payments. Indeed, whatever the actual shape of the repayment schedule involved, our debt-based money creation methodology chains our entire productive and creative potential to calendar time. (Papazian 2022, 214)

Irrespective of the length of time involved, or how many times interest is compounded, a calendar time-based conceptualisation of money creation instruments chains governments, government agencies, municipalities, small businesses, households, individuals, corporations, and even banks to calendar time payments.

I must emphasise that in the context of debts and banking, we are talking about 'calendar time.' The nature of time is a debateable subject from a theoretical physics perspective, and economists have also considered the relevance of psychological time to the performance of investments. What lies at the heart of our monetary system and what is used and applied by banks and central banks in debt transactions is simply calendar time.

Calendar time is a human construct. The Gregorian calendar, the one we use across the world, was designed by Aloysius Lilius (c. 1510–1576), also known as Luigi Lilio, and was introduced by Pope Gregory XIII in October 1582. Four hundred years later, in 1982, the Vatican organised a conference commemorating the 400th anniversary of the calendar. In the proceedings of the conference (Coyne et al. 1983), we can read the following:

> In most chronology, or time reckoning, the basic unit is the day. Time reckoning is all about counting an unbroken succession of days by making them up in various appropriate bundles. Some of the bundles, such as the week, are artificial and comprise integral numbers of days. Their use presents no problem more complex than counting inches in sets of 36, called yards. But other time intervals used in calendars, notably months and years, are based – like the days themselves – on astronomical periods. (Coyne et al. 1983, 3)

It is particularly interesting to note that the core substance of our calendar has astronomical roots. The basic unit of our calendar, a day, measures the rotation of Earth on itself, and a year is the revolution of the Earth around the sun. While calendar time is founded on observations of our planet's movements in outer space, relative to itself and the sun, when used as a central pillar of our monetary architecture, it limits our ability to create and invest the resources we need to expand our reach in outer space.

8.3   A Muzzle in Space

How is this relevant and why, most importantly, does calendar time act as a muzzle on our spatial potential when used as a foundational pillar of money creation?

Calendar time is a human invention that allows us to structure and navigate our productive life on the planet and it is a central pillar of the world economy. As such, this discussion does not negate this fact, and it does not argue, in any shape or form, that we should stop using the calendar. The argument here concerns the use of calendar time as the central pillar of the instruments we use for the creation of money, i.e., debts.

In Chapter 3, discussing the layered conceptualisation of space, I introduced the Prime Meridian, one of the conceptual foundations through which we map and structure our terrestrial space. Quoting Withers (2017) once again: "[t]he Prime Meridian is the line and the point at which the world's longitude is set at 0°. It does not exist in any strict material sense, yet through maps and clocks, the prime meridian governs the life of every human on Earth" (Withers 2017, 5).

The history of the Prime Meridian is fascinating. There was a time in human history when we had more than one 'prime meridian' used by different empires. Recognising the challenge of coordinated communication and trade based on different reference points, the Greenwich Prime Meridian was selected in 1884. A hundred years later, in 1984, and by international agreement, the Prime Merdian was replaced by the International Reference Meridian (IRM) based on the International Earth Rotation and Reference Systems Service (IERS), the international body that maintains global time and reference standards.

The reason the Prime Meridian (Fig. 8.2) is relevant to this discussion is because it helps us define our progression in calendar time. This happens thanks to the parallel and linked concept of the International Date Line (IDL), located halfway around the world from the Prime Meridian.

The international date line, established in 1884, passes through the mid-Pacific Ocean and roughly follows a 180 degrees longitude north-south line on the Earth. It is located halfway around the world from the prime meridian — the 0 degrees longitude line in Greenwich, England… The international date line functions as a "line of demarcation" separating two consecutive calendar dates.

When you cross the date line, you become a time traveler of sorts! Cross to the west and it's one day later; cross back and you've "gone back in time… Despite its name, the international date line has no legal international status and countries are free to choose the dates that they observe. While the date line generally runs north to south from pole to pole, it zigzags around political borders such as eastern Russia and Alaska's Aleutian Islands. (NOAA 2024)

To understand where the limitation comes into play, think of all the instruments used to create money on Earth—loans given by commercial banks, bonds purchased by central banks. They all have maturity dates that depend on the turning of the grid in Fig. 8.2. As the grid (Earth) turns 365.242 times, the calendar moves a year. The grid acts as a muzzle because the entire money supply created within it is attached to the *fixed paced turn of the grid* and the fixed revolution around the sun.

It is indeed ironic that our calendar, which is built upon Earth's movements in outer space, rotation on itself and revolution around the sun, limits our ability to create and invest the resources we need to expand in outer space. This is so given the fixed pace of these movements relative to the enormous landscape within which other celestial bodies have very different orbits.

When we link money creation to calendar time, we limit our ability to invest in space timelessly. In other words, the entire grid depicted in Fig. 8.2, starting from the Prime Meridian, which also defines the International Date Line and our time zones, acts like a muzzle—we cannot create money that does not depend in one way or another to the rotation and revolution of

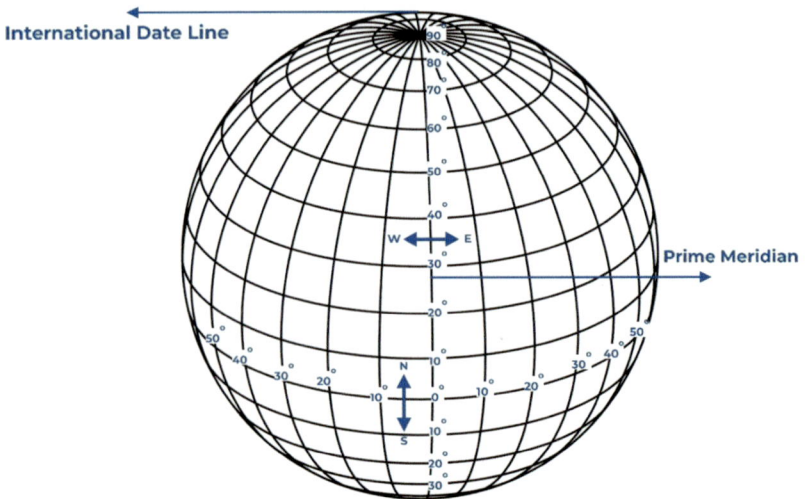

Fig. 8.2 Prime meridian and international date line (*Source* Author)

Earth. Figure 8.3 depicts the tropical orbits of the planets in our solar system, including Pluto's Sidereal orbit, and identifies the number of Earth days they take to orbit the sun. As you can see, the planets' revolution around the sun is specific to each planet. Given distance from the sun and speed, celestial bodies have different 'years.'

Table 8.1 expands on these differences to demonstrate that a calendar year on Earth, or one revolution around the sun, to which our money creating instruments are linked, is a special variable particular to our planet. Mars takes roughly 687 Earth days and Jupiter takes roughly 4333 Earth days to orbit the sun. In other words, when our monetary system is linked to our 365.242-day tropical orbit and calendar, it chains us to a measurement that does not match the specifics of the other planets we intend to settle on.[2]

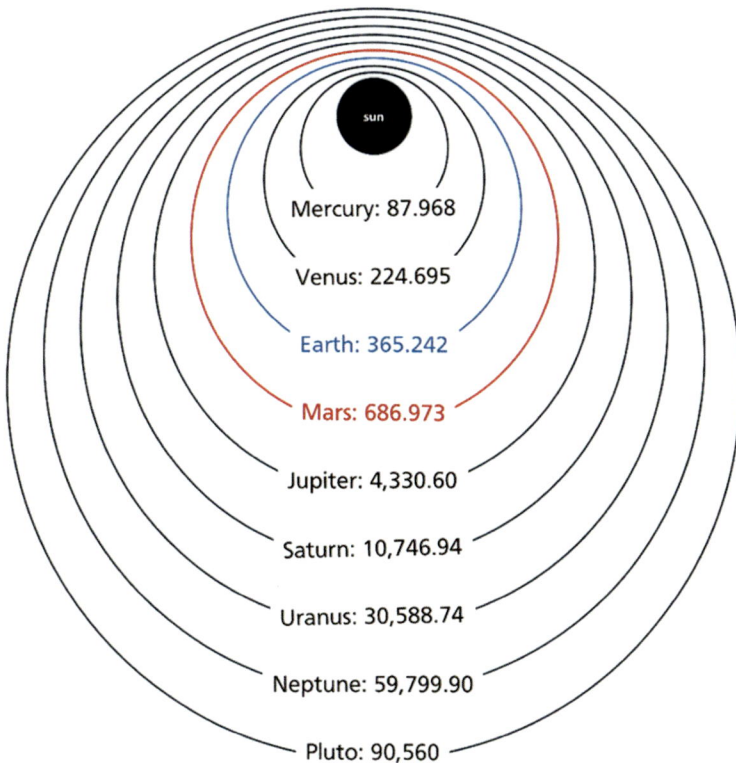

Fig. 8.3 Planets' tropical and Pluto's sidereal orbits in earth days (*Source* Author using NASA [2024a, b, c, d, e, f, g, h, i])

----

[2] A year is the revolution of Earth around the sun, or one full orbit, and it is often measured in three different ways. A tropical year, which our calendar strives to follow measures the time required for the sun to return to the same position, to pass from vernal equinox to vernal equinox. It is equal

Table 8.1 Planetary orbits around the sun in earth days

	Sidereal orbit around the sun in earth days	Tropical orbit around the sun in earth days	Sidereal one year on earth equivalent on planet	Sidereal one year on planet equivalent on earth
Mercury	87.969	87.968	4.152099	0.240842
Venus	224.701	224.695	1.62552	0.615188
Earth	365.256	365.2422	1	1
Mars	686.98	686.973	0.531684	1.880818
Jupiter	4,332.59	4,330.60	0.084304	11.86179
Saturn	10,759.22	10,746.94	0.033948	29.45666
Uranus	30,685.40	30,588.74	0.011903	84.01067
Neptune	60,190	59,799.90	0.006068	164.7885
Pluto	90,560	Unknown	0.004033	247.9357

Source Compiled by Author using NASA (2024a, b, c, d, e, f, g, h, i)

In Table 8.2, you can see the minimum and maximum distances of these planets from Earth and the time it would take for light, at a speed of 299,792,458 m/s, and the fastest human object, NASA's Parker Probe at a speed of 176,462.78 m/s (NASA 2024), to travel to these planets. Naturally, these are hypothetical cases.

The point here is that even at the speed of light, and at the speed of the Parker Probe, our calendar will still be out of sync. With current human speeds, the difference is much more significant, and given that their orbits are not similar to Earth's orbit, our calendar time dependent monetary architecture robs us of the opportunity to create money independently of our orbit, prevents us from creating money relative to the vastness of space rather than the fixed and limited movements of Earth. This is particularly relevant when we intend to change orbit.

This is not about general relativity and time dilation, but the simple fact that using calendar time as a foundational pillar of money creating instruments chains us to the rotation and revolution of Earth, which should not come into play when creating resources to invest in a trip to Mars or elsewhere in the solar system. Indeed, if the plan is to leave Earth and establish new habitats and/or terraform a planet that takes almost twice the time to orbit the same sun, linking our monetary and investment capability to the rotation and revolution of Earth is particularly awkward.

---

to 365.2422 days. The sidereal year, which is 365.256 days, slightly longer, measures the revolution based on the time it takes for Earth to return to the same place in its orbit, in relation to the stars. The anomalistic year, based on the elliptical nature of Earth's orbit, which measures the time it takes for Earth to return to its closest distance point to the sun, called perihelion.

Table 8.2 Planetary distances and travel time

	Min distance from earth (Million Km)	Max distance from earth (Million Km)	Time light travels from earth to planet at min (Seconds)	Time light travels from earth to planet at max (Seconds)	Time Parker probe travels from earth to planet at min (Seconds)	Time Parker probe travels from earth to planet at max (Seconds)
Mercury	77.3	221.9	257.85	740.18	438,052.72	1,257,488.97
Venus	38.2	261	127.42	870.60	216,476.25	1,479,065.44
Earth	0	0	0.00	0.00	0.00	0.00
Mars	54.6	401.4	182.13	1,338.93	309,413.69	2,274,700.65
Jupiter	588.5	968.5	1,963.02	3,230.57	3,334,980.90	5,488,409.51
Saturn	1204.28	1652.48	4,017.05	5,512.08	6,824,555.30	9,364,467.68
Uranus	2580.6	3153.5	8,607.96	10,518.94	14,624,047.07	17,870,624.05
Neptune	4319	4711	14,406.63	15,714.20	24,475,416.29	26,696,847.91
Pluto	4284.7	7528	14,292.22	25,110.71	24,281,041.02	42,660,554.25

Source Compiled by Author using NASA (2024a, b, c, d, e, f, g, h, i)

While central to structuring communications and transportation, and critical for navigation and positioning, the prime meridian, our longitudes and latitudes, our calendar, and our time zones end up acting like a limiting muzzle when used as the core structural pillar of our monetary architecture.

The calendar time-based conceptualisation of money creation chains our productive horizon and chains our reach to the steady turn of the planet on itself and revolution around the sun. Our ability to reach far beyond Earth in terms of distance/time and create life and habitats outside of the grid (Fig. 8.1) becomes a struggle.

Calendar time acts as a muzzle for private and public entities in the outer space sector. From a public perspective, yearly budgets, government debts, and subsequent monetary obligations are the structural hook where calendar time latches on to our investment and productive potential in outer space. In other words, that is where the muzzle is attached to our ability to create and invest what we need in order to explore and expand in outer space.

In a recent article in Science News, Mann (2024) writes:

NASA's 2024 budget comes to $24.875 billion, a 2 percent cut relative to last year and 8.5 percent less than the requested funding. That's the biggest discrepancy between requested and appropriated funding for the agency since 1992.

The budget's approval immediately left it up to NASA administrators to figure out how to adapt and cover the $509 million gap.

"We know we are definitely in an imperfect environment, and we acknowledge this is a very challenging time," Nicola Fox, NASA's associate administrator for science, said in March during a public town hall. But, she vowed, NASA "will use every single penny to do great science. (Mann 2024)

The imperfect environment is exactly that, fundamentally constrained by the monetary architecture we have created for ourselves, in the middle of a vast cosmic landscape. On top of the structural impediments created through the very architecture of our monetary system, reporting requirements are another hook through which calendar time prevents our timeless expansion in outer space.

Indeed, accounting and reporting affect both, private and public entities. Anyone who has been involved in any kind of public and/or corporate reporting is well aware of the bureaucratic time-consuming nature of the process. This is not to argue against accountability, not at all, but about the calendar time-based yearly reporting process that chains much of our time and processes to the revolution of the Earth around the sun. A feature of our economies that is particularly irrelevant and prohibitive when it comes to outer space projects, where our targets and objectives work on entirely different calendars and involve immense distances out of Earth's orbit.

8.4   Conclusion

Calendar time is a fixed and limited concept based on the revolution and rotation of Earth. Why does our monetary architecture have to be linked to the rotation and revolution of Earth? Have you ever asked yourself that question? While Earth in space is within a vast and unlimited context, we have decided to build our monetary architecture on the unchanging fixed pace of our planet's revolution around the sun. The limited calendar may very well be considered the 'limited resources' economists often refer to. Indeed, such a constraining definition could only have come about in the absence and omission of space.

The use of calendar time in the organisation and structuring of our days and our productive activities is obvious, and not a subject of debate. However, its use and function as a sole organising principle of our monetary architecture is questionable, and a muzzle on our ability to (timelessly) create and invest the resources we need for our expansion in outer space.

This structural feature of our spaceless monetary architecture imposes severe limitations. Indeed, when our monetary architecture is chained to the revolution of our planet around the sun and rotation on itself, we should not be surprised by our inability to invest and expand our reach in outer space. When the whole point of the exercise is to go beyond Earth, the fact that our resources are chained to the rotation and revolution of Earth is at best a bottleneck, and at worst, a debilitating evolutionary misconception.

Indeed, debt-based money chains us to the surface of the planet, through our own conceptual projections that underpin our entire monetary system.

References

Allais, M. 1972. Forgetfulness and Interest. *Journal of Money, Credit and Banking* 4(1), 40–73. https://doi.org/10.2307/1991402. Accessed 2 March 2022.

Allais, M. 1974. The Psychological Rate of Interest. *Journal of Money, Credit and Banking* 6(3), 285–331. https://doi.org/10.2307/1991172. Accessed 2 March 2022.

Bank of England. 2021. Bank of England Corporate Bond Purchase Scheme: Eligible Bonds List. Bank of England. https://www.bankofengland.co.uk/-/media/boe/files/markets/corporate-bond-purchases/bonds-eligible-for-the-corporate-bond-purchase-scheme.xlsx. Accessed 5 June 2024.

Coyne, G.V., Hoskin, M.A., Pederson, O. 1983. Georgian Reform of the Calendar. Proceedings of the Vatican Conference to Commemorate Its 400th Anniversary. Specola Vaticana. https://www.pas.va/content/dam/casinapioiv/pas/pdf-volumi/extra-series/es3pas.pdf. Accessed 12 March 2024.

NASA. 2024. *NASA's Parker Solar Probe Completes 18th Close Approach to the Sun.* NASA Blog. https://blogs.nasa.gov/parkersolarprobe/2024/01/08/nasas-parker-solar-probe-completes-18th-close-approach-to-the-sun/. Accessed 12 February 2024.

NASA-SSDCA. 2024a. Mercury Fact Sheet. NASA Space Science Coordinated Archive. https://nssdc.gsfc.nasa.gov/planetary/factsheet/mercuryfact.html. Accessed 12 May 2024.

NASA-SSDCA. 2024b. Venus Fact Sheet. NASA Space Science Coordinated Archive. https://nssdc.gsfc.nasa.gov/planetary/factsheet/venusfact.html. Accessed 12 May 2024.

NASA-SSDCA. 2024c. Earth Fact Sheet. NASA Space Science Coordinated Archive. https://nssdc.gsfc.nasa.gov/planetary/factsheet/earthfact.html. Accessed 12 May 2024.

NASA-SSDCA. 2024d. Mars Fact Sheet. NASA Space Science Coordinated Archive. https://nssdc.gsfc.nasa.gov/planetary/factsheet/marsfact.html. Accessed 12 May 2024.

NASA-SSDCA. 2024e. Jupiter Fact Sheet. NASA Space Science Coordinated Archive. https://nssdc.gsfc.nasa.gov/planetary/factsheet/jupiterfact.html. Accessed 12 May 2024.

NASA-SSDCA. 2024g. Saturn Fact Sheet. NASA Space Science Coordinated Archive. https://nssdc.gsfc.nasa.gov/planetary/factsheet/saturnfact.html. Accessed 12 May 2024.

NASA-SSDCA. 2024f. Uranus Fact Sheet. NASA Space Science Coordinated Archive. https://nssdc.gsfc.nasa.gov/planetary/factsheet/uranusfact.html. Accessed 12 May 2024.

NASA-SSDCA. 2024h. Neptune Fact Sheet. NASA Space Science Coordinated Archive. https://nssdc.gsfc.nasa.gov/planetary/factsheet/neptunefact.html. Accessed 12 May 2024.

NASA-SSDCA. 2024i. Pluto Fact Sheet. NASA Space Science Coordinated Archive. https://nssdc.gsfc.nasa.gov/planetary/factsheet/plutofact.html. Accessed 12 May 2024.

NOAA. 2024. What Is the International Date Line? National Oceanic and Atmospheric Administration. https://oceanservice.noaa.gov/facts/international-date-line.html. Accessed 12 May 2024.

Papazian, A. 2022. *The Space Value of Money: Rethinking Finance Beyond Risk and Time.* New York: Palgrave Macmillan. https://doi.org/10.1057/978-1-137-59489-1.

Papazian, A. 2023. *Hardwiring Sustainability into Financial Mathematics: Implications for Money Mechanics.* New York: Palgrave Macmillan. https://doi.org/10.1007/978-3-031-45689-3.

Withers, W.J.C. 2017. *Zero Degrees: Geographies of the Prime Meridian.* Harvard University Press.

9

Monetary Gravity: A Leash in Space

Exploration is really the essence of the human spirit.
Frank Borman, Gemini 7 and Apollo 8 Astronaut, 1968

Life in orbit is spectacular.
Tim Peake, Soyuz TMA-19M Astronaut, 2015

The next challenge with debt-based money is directly linked to the use of calendar time as a foundational element of money creation instruments, but it goes one step further. All debt instruments used by central and commercial banks require the repayment of principal and interest to the original source, the creator of money. In the case of central banks, their credit facilities and purchased bonds must continuously pay interest and principal. That is what makes them eligible instruments for monetary policy purposes. The same applies to commercial banks. Making loans and creating deposits is built on the reliability of the borrower and the repayment of principal and interest. These obligations to repay are defined by calendar time intervals.

This chapter is dedicated to demonstrating that the calendar time-based obligation to repay the money creators acts as a leash in space. This is so because it defines the hypothetical limit on the distance we can travel in space before we have to return to some bank.

© The Author(s), under exclusive license to Springer Nature
Switzerland AG 2024
A. V. Papazian, *Financing the Race to Space*,
https://doi.org/10.1007/978-3-031-73102-0_9

9.1   Monetary Gravity

Monetary gravity describes the backward loop to the creator of money imposed through debt instruments—an artificially created force that acts like a leash (Papazian 2022). This is true for all kinds of debt instruments, and it remains true even when debts can be rolled over and postponed. It also remains true when we can repay electronically without physically going back to the bank.

The easiest way to understand this is through the same cashflow timeline we discussed in chapter eight. This time, however, instead of looking at the timeline through the eyes of the investor, we will look at it from the eyes or experience of the debtor (Fig. 9.1). Drawing the cashflows from the perspective of the debtor, as in Fig. 9.1, we can see that the repayment of principal and interest forces the borrower to return to the bank at regular calendar time intervals. The bond valuation equation (Eq. 9.1) assesses the cash flows that the debtor, upon borrowing, will be paying to the bondholder (the coupon payments (C_t) and the Par Value (P) or Principal Amount at maturity).

Naturally, this logic applies to all loans, and bonds can be owed to anybody, not just banks. The main point here is that calendar time-based repayment obligations, the foundational components of our monetary architecture, act as a leash on all involved.

$$\text{Bond Value} = \sum_{t=1}^{n} \frac{C_t}{(1+r)^t} + \frac{P}{(1+r)^n} \tag{9.1}$$

n = Time to Maturity
t = Moving time
r = Discount Rate or Yield to Maturity
P = Par Value of Bond
C_t = Coupon Payments

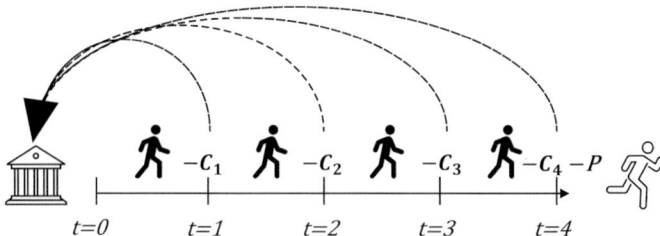

Fig. 9.1 Cash flow timeline debtors' perspective (*Source* Author)

This is not a moral judgement on debt. I am describing the mechanics of the instruments. I am also not suggesting that loans and debts should not be repaid. That is not the argument here. The point is to understand the limitations that a debt-based money creation methodology imposes on us, on our resources, investments, and productive potential.

9.2 A Leash in Space

Debt-based money acts as a leash on our species because the obligation to repay the creator of money at different calendar time intervals limits the hypothetical distance we can travel in space before we have to return to some bank. Of course, with electronic banking, we can now transfer debt payments without physically returning to the bank. While this is true, it does not eliminate the calendar time linked obligation to consider the money creator, and it does not remove pressure to generate the necessary cash flows before payments are due. Even when debts can be rolled over, or refinanced, the structural features of debt-based money impose a limit on how far in space a process can go before having to consider its obligation to some bank, central or commercial.

I use this example to illustrate the level of the absurd. Using the below conceptual equation (Eq. 9.2), and assuming uniform terrestrial conditions for simplicity and the purpose of the argument, we can calculate the limit on distance travelled imposed by a debt instrument requiring a monthly interest payment. Table 9.1 provides the limits on distances given the speed of Usain Bolt, Koenigsegg Jesko Absolut, the Parker Probe (NASA 2023), and light.

$$\text{Maximum Distance}_{\text{Light}} = \text{Speed in} \frac{m}{s} \times \text{Time Interval in } s \qquad (9.2)$$

The distance light can travel in one month, also known as a light-month, is the distance that light travels in an absolute vacuum in one full month. The speed of light is equal to 299,792,458 m/s. Assuming 30 days in a month,

Table 9.1 Distance travelled in a month in metres

	Usain Bolt	Koenigsegg Jesko Absolut	Parker Solar Probe	Light
Distance (m)	27,060,480	382,379,616	457,391,520,000	777,062,051,136,000

Source Author, updated from Papazian (2022)

and 86,400 seconds in each, in one month light travels 777,062,051,136,000 metres, which is equivalent to approximately 777 Tm (1 Terametre = 1,000,000,000,000 metres). At that point, or slightly earlier, light will have to make arrangements for a wire transfer to the bank.

Usain Bolt, the Jamaican sprinter, set the world record in 2009 in the 100-metre sprint at 9.58 seconds, giving him a speed of 10.44 metres per second, which means the furthest Bolt can run in one month is 27,060,480 metres. One of the fastest production cars, the Koenigsegg Jesko Absolut, is reported to have a speed of 330 miles per hour, or 147.523 metres per second, the furthest the Koenigsegg Jesko Absolut can travel in a month is 382,379,616 metres. NASA's Parker Solar Probe (NASA 2023) achieved speeds of 635,266 km per hour or 176,462.78 metres per second, the furthest it can travel within a month is 457,391,520,000 metres.

These hypothetical limits and examples demonstrate that we, and our fastest tools and inventions, and light itself, will experience a limit on distance travelled given the obligation to return to some bank in an interval of calendar time.

When we impose a structural condition to repay the created money given a certain number of rotations or revolutions of Earth, we are in fact chaining a large segment of society and our productive and business activities to banks and central banks, to the surface of this planet, to its rotation and revolution. It is very important to clarify that this leash exists not just for those outer space projects that are financed directly through debts. This is so because the entire money supply creation process is built on debts. The private and public outer space sectors are affected by this fact.

The private outer space sector, as described in chapter two, is Earthbound through the very necessity to chase and deliver profits, which can only be achieved by pursuing Earthly money supply, which is continuously created through debt. While the funding sources of the private outer space sector are relatively more diverse, using revenues, debts, and equity, it is still directly and indirectly chained to calendar time, and the debt obligations that underpin the money it chases.

The public outer space sector is directly constrained by the debt-based nature of money because public governmental spending is linked to taxes and debts. As such, the levels of public debt have a direct impact on how much new debts governments can issue, while government bonds are used to back the issuance of currency and the creation of central bank reserves. Moreover, as discussed in chapter seven, bonds used for monetary policy purposes must be investment grade with secure cash flows. Which means that the issuing entity must always consider the levels of risk it takes on. In other words, this leash is in the background, and it has implications on how indebted entities like households, businesses, corporates, and governments think and behave, and has indirect implications for those who are not in debt but must pursue the money that is created through debt.

Naturally, this leash becomes more obvious when we consider the simple case of funding outer space investments directly through debt. Indeed, an investment that aims to build and deploy a Mars habitat is not only dealing with great distances, but also an entirely different orbit, an entirely different calendar. As discussed in the previous chapter, these immense distances on their own create structural challenges for debt financing.

Using debt, and only debt, as the main logic of money creation constrains our ability to create and invest the resources we need to expand in outer space. Using calendar time as a central pillar and imposing a backward loop to the creator of money limit our ability to travel long distances without looking back. Exploring outer space is exactly about going beyond Earth, and our planet's revolution around the sun should not limit how much we create and invest in achieving that objective.

However, this is only one side of the story. Every leash has two ends and there is usually an entity or entities holding the leash. Ironically, in humanity's case, humanity is at both ends. We have chained ourselves to this system. Our architecture is maintained by our belief in and service to this architecture. Naturally, different parts of humanity may find themselves at either end of this leash. Ultimately, however, we are all chained to calendar time, to debt obligations that are invented into being through our monetary architecture, upheld by our legal systems, and enforced by governments who are themselves in debt. This leash and the muzzle are self-inflicted.

As such, the leash holds us all back, even if within a specific time window, some of us feel better off because they are holding the leash, rather than being held back by the leash. Ultimately, the fact that our monetary architecture is built on debt, on calendar time-based debt obligations, and these obligations are themselves dependent on credit ratings, and the necessity to maintain them, we are constrained by what affects those ratings. In other words, given that high risks and distant returns may affect our ratings negatively, our architecture reinforces the bias in our financial value framework and ensures the leash is tight and securely hooked on.

9.3   Conclusion

Using calendar time as a foundational pillar of debt-based money acts as a muzzle on our ability to invest in space timelessly. Calendar time-based obligations to repay the creators of money act as a leash on our ability to conquer great distances without having to look back, or return to some bank, or cut budgets to appease mortal investors worried about risk and time.

> Monetary gravity is far more constraining than actual gravity, where our propulsion technologies have already taken us to the moon and Mars. Monetary gravity chains us to the surface of the planet, spinning around an imaginary calendar, while ignoring the vast landscape of space and the resources within. We are instead forced to deal with limited budgets, credit ratings, and a host of constraints like a debt ceiling in the US (Papazian 2022, 218)

Moreover, the leash has implications even if we allow ourselves to extend the leash continuously. This is the case, for example, of the US Debt Limit or Ceiling, which I discuss in chapter fifteen. Just because we can kick the can down the road, roll over debt, or refinance, it does not mean that we are free from the limitations imposed by the architecture. In fact, quite the opposite. Extending the leash is a testimony of the fact that there is a leash that needs extending.

Given that debts are an important source of funding for public expenditures, along with taxes, a debt-based architecture imposes limitations on how much we can spend given existing debt levels and monetary conditions. This is particularly constraining given that the public outer space sector is the only segment that does not seek monetary rewards or profits, and thus it is theoretically not bound to chase Earthly money supply.

The calendar time obligations to repay the money creators impose a limitation on the distance we can travel in space before having to return to some bank. This leash chains us to banks, to Earthly money supply, and the rating and legislative paraphernalia through which debts are monitored and enforced.

In an expanding universe where we have an entire galaxy to explore, we are leashed to imaginary debts, to the surface of this planet, to the constant and fixed paced rotation and revolution of Earth.

References

NASA. 2023. For the Record: Parker Solar Probe Sets Distance, Speed Marks on 17th Swing by the Sun. NASA. https://blogs.nasa.gov/parkersolarprobe/2023/09/28/for-the-record-parker-solar-probe-sets-distance-speed-marks-on-17th-swing-by-the-sun/. Accessed 12 April 2024.

Papazian, A. 2022. *The Space Value of Money: Rethinking Finance Beyond Risk and Time*. New York: Palgrave Macmillan. https://doi.org/10.1057/978-1-137-59489-1.

Papazian, A. 2023. *Hardwiring Sustainability into Financial Mathematics: Implications for Money Mechanics*. New York: Palgrave Macmillan. https://doi.org/10.1007/978-3-031-45689-3.

10

Monetary Hunger: A Whip in Space

I could see no border on earth from space.
Mamoru Mohri, STS-47 Endeavor Astronaut, 1992

Just like our beautiful Planet, we are all insignificant compared to the vastness of the Universe. We are all essentially equal. We are not alone and isolated: when you hurt someone, you hurt yourself; when you help someone, you help yourself.
Marcos Pontes, Soyuz TMA-8 Astronaut, 2006

In Chapter 8 we observed how an apparently benign feature of debt, i.e., calendar time, acts as a muzzle in space. In chapter nine, we saw how calendar time-based debt obligations act as a leash in space. The third systemic challenge imposed by debt-based money is what I call *monetary hunger*, which, coupled with our laws and regulations governing default, acts as a whip in space.

According to the latest May 2024 report published by the Institute for International Finance (IIF), "Global debt rose by some $1.3 trillion to a new record high of $315 trillion in Q1 2024" (IIF 2024). The report reveals that this increase is the second consecutive quarterly rise and is driven by emerging markets, where debt levels rose to over $105 trillion. Two-thirds of the $315 trillion is in mature economies, where Japan and the United States are the largest contributors.

© The Author(s), under exclusive license to Springer Nature
Switzerland AG 2024
A. V. Papazian, *Financing the Race to Space*,
https://doi.org/10.1007/978-3-031-73102-0_10

Global and national debt figures are directly linked to the fact that money is continuously created via debt, which is a fundamental driver of money supply growth in our current architecture.

Whatever the concocted logic behind the use of debt as a central pillar of money creation, the limitations it imposes on us, on a species in space, are seriously prohibitive. Indeed, as discussed, debt-based money has evolutionary implications. It plays a central role in undermining our ability to create and invest the resources we need to secure sustainability on Earth and drive our expansion in outer space.

10.1 Monetary Hunger

In any debt-based economy, and at any point in time, irrespective of past or current capital accumulation, a large segment of society, including households, municipalities, governments, corporations, and banks, is chasing available cash and deposits to pay calendar time-linked debt obligations. Debt-based money creates this chase, this monetary hunger, in any debt-based economy irrespective of the actual levels of debt to GDP ratios. This is so given that money is continuously created via debt. (Papazian 2023, 122)

Irrespective of how much capital accumulation is achieved within a debt-based economy, the creation of new money through new debt ensures that this monetary hunger is a permanent feature of our economic reality. The necessity to chase cash and deposits to fulfil debt obligations, even though taken for granted as an integral component of our current system, creates this artificial hunger.

To contextualise this discussion further, Chart 10.1 depicts total outstanding public and private debts in the United States, between 2000 and 2022, growing from 28.63 to 93.50 trillion US Dollars.

Looking at IMF historical world debt data, Chart 10.2, we can see the long-term trend of debt to GDP percentage increases. In 2022, world public debts represented 92.4% of world GDP and private debts amounted to 145.7% of world GDP, while global total debt stood at 238.1%. The relative decline in 2020–2022 is mainly due to the global pandemic. As mentioned in the introduction, debt levels have already registered their second consecutive quarterly rise by Q1 2024.

In other words, despite and in parallel to capital accumulation and output growth, and irrespective of who and how many are accumulating wealth, our entire monetary architecture is built on public and private debts.

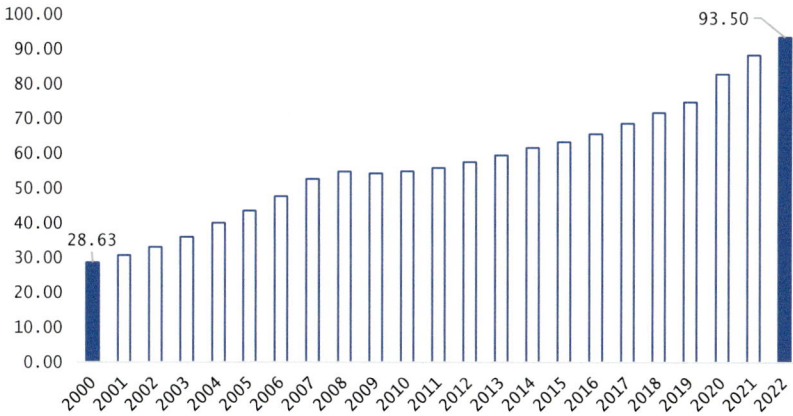

Chart 10.1   Total Outstanding Public and Private Debt USA, 2000–2022 in trillion USD (*Source* Statista 2023)

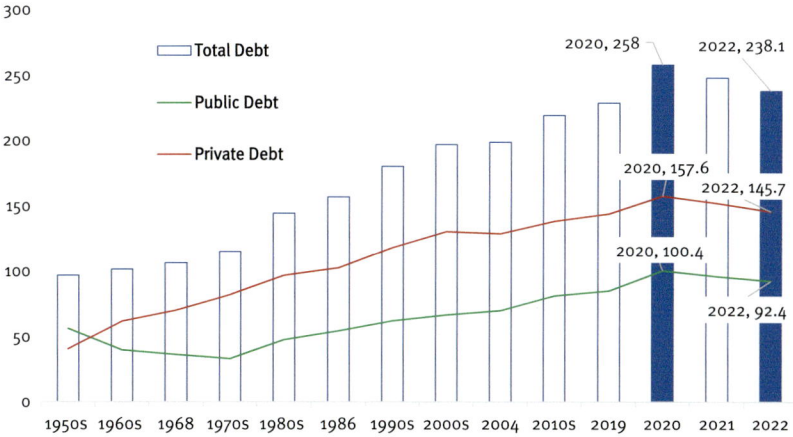

Chart 10.2   World Total Debt, Public Debt, and Private Debt as % of World GDP (*Source* IMF 2023)

10.2   A Whip in Space

Have you ever encountered a debt instrument or transaction where the payments are voluntary? Optional? Obviously not. Paying back our debt instalments on time, as per calendar time, is a legally binding commitment that can affect our credit worthiness, our ability to raise new funds, and may even lead to the loss of collateral assets. In other words, the artificially created monetary hunger, coupled with the legal system that enforces it, is an effective system that mimics a whip in space.

As such, the monetary hunger created through debt-based money, and the tools that are associated with the monitoring and enforcement of debts, define the behaviour and actions of economic agents. The artificial hunger is enforced with a real whip that can destroy businesses and leave families homeless.

In contrast, our emissions reporting and sustainability standards are still voluntary. Despite the rhetorical and bureaucratic frenzy surrounding sustainability, reporting our impact on space is still an optional process. To put it mildly, when our sustainability standards are optional and our debt payments are not, given the threat of default and the absence of space and space responsibility from our financial value framework, everyone will serve their debts before the environment or space (outer space included). Our debt-based monetary architecture has been given primacy over our ecosystem, over space in general.

There is of course a difference between public and private debts. They create their own kind of monetary hunger and the nature of the whip differs in each case. Whether business, corporate, household, consumer, or other, private borrowers are far more exposed to the whip than public borrowers. This does not mean that the whip does not have a direct impact on governments and their budgets. It is just a different kind of whip. While public borrowers cannot lose collateral assets the way private borrowers may, both public and private borrowers can lose their credit ratings. This has a direct impact on the cost of borrowing or cost of capital. Moreover, governments can also default on their debts with economy wide consequences.

When governments are in trouble, they get downgraded, and they may be cut off from markets. This can lead to economy wide challenges, socioeconomic and political turmoil, budgetary austerity, and significant reduction in public services. Even if governments can theoretically issue new debts and roll over old ones continuously, their credit ratings and stability have serious and far-reaching implications that constrain this ability. To substantiate these observations, the below discussion is a brief snapshot of what happened in Greece following the 2007/2008 financial crisis and the subsequent sovereign debt crisis.

Chart 10.3 depicts Greece Credit Default Swap prices. Credit Default Swaps are derivative instruments used as a form of insurance. "A credit default swap (CDS) is a contract between two parties in which one party purchases protection from another party against losses from the default of a borrower for a defined period of time" (CFA Institute 2024). CDSs reflect market sentiment vis-à-vis a specific borrower and its debts, which affects its access to the market as well as its cost of borrowing.

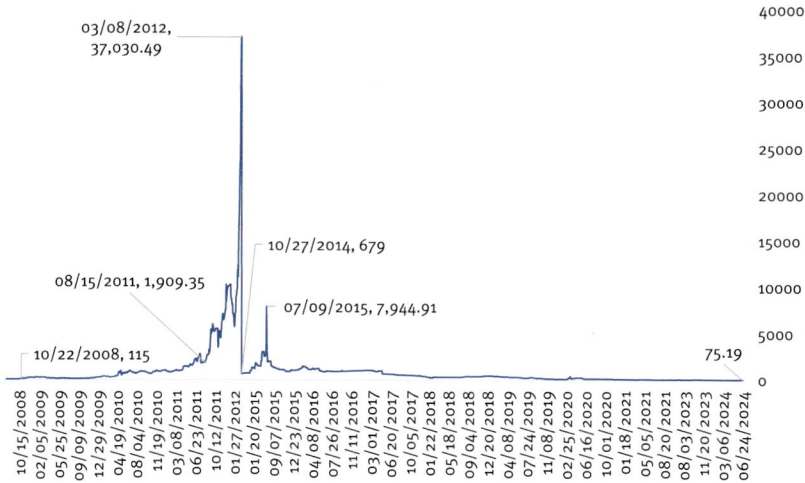

Chart 10.3   Greece Credit Default Swaps (CDS) 5 Years USD Prices (*Source* Investing 2024)

Chart 10.4 depicts the 3-year Greek government bond prices and yields. Mirroring Chart 10.2 and revealing the crisis moment and the debt restructuring of 2012 (prices are per €100 nominal, and yields are in percentages).

> The 2012 Greek debt exchange and subsequent buyback was a key episode in the Eurozone debt crisis. It was the largest debt restructuring in the history of sovereign defaults, and the first within the Eurozone. Though it achieved historically unprecedented debt relief – amounting to 66% of GDP – it was 'too little too late' in terms of restoring Greece's debt sustainability. (Xafa 2014)

Indeed, in February 2012, the European Central Bank (ECB) had already suspended the eligibility and use of Greek bonds for monetary policy purposes. A key feature of the Quantitative Easing strategy discussed in previous chapters, and also implemented by the ECB (ECB 2012).

On the 9th of March 2012, Moody's published an assessment of Greece's debt restructuring and exchange, classifying it as a default.

> Moody's Investors Service says that it considers Greece (C/no outlook) to have defaulted per Moody's default definitions further to the conclusion of an exchange of EUR177 billion of Greece's debt.... Moody's understands that 85.8% of debtholders holding Greek-law bonds issued by the sovereign have agreed to the exchange, with the vast majority of remaining bondholders likely to be drawn in following the exercise of Collective Action Clauses that will be inserted pursuant to a recent Act by the Greek parliament. The terms of

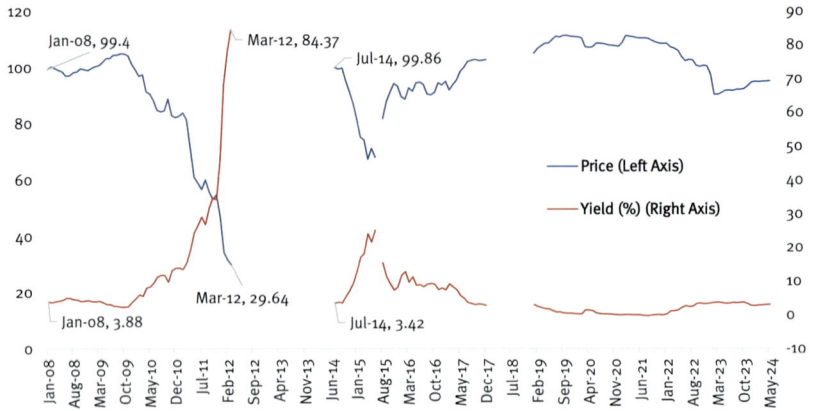

Chart 10.4 Greek Government benchmark 3-year bond prices and yields (*Source* Bank of Greece 2024)

> the exchange entail a discount -- a loss to creditors -- of at least 70% on the net present value of existing debt.... According to Moody's definitions, this exchange represents a 'distressed exchange', and therefore a debt default. (Moody's 2012)

All these developments led to Greece losing its access to capital markets, and the imposition of radical policy reforms by the new lenders bailing it out, i.e., the European Central Bank (ECB), the International Monetary Fund (IMF), and the European Financial Stability Facility (EFSF).

Naturally, the downgrading process began much earlier. Table 10.1 lists Moody's Rating announcements and actions for Greece from 2002 to 2023. While very often these ratings are late to reflect market realities, they nevertheless have their impact on costs of borrowing. Indeed, the authenticity and integrity of these ratings are often under scrutiny. Bear Stearns and Lehman Brothers had investment grade ratings until five days before their collapse in 2008.

The above discussion is not about absolving defaulters, or justifying failed policies that lead to defaults. It is about recognising that the artificially created hunger, based on public and private debts, and the associated regulatory and legal framework, function as a whip in space. For private borrowers the primary threat is a loss of assets, for public borrowers the primary threat is loss of market access and policy control.

Table 10.1 Moody's rating updates and actions on Greece, 2002–2023

Date	Moody's rating, outlook
September 15, 2023	Ba1, stable
March 17, 2023	Ba3, positive
November 6, 2020	Ba3, stable
March 1, 2019	B1, stable
February 21, 2018	B3, positive
June 23, 2017	Caa2, positive
September 25, 2015	Caa3, stable
July 1, 2015	Caa3, negative watch
April 29, 2015	Caa2, negative
February 6, 2015	Caa1, negative watch
August 1, 2014	Caa1, stable
November 29, 2013	Caa3, stable
March 2, 2012	C, negative
July 25, 2011	Ca, negative
June 1, 2011	Caa1, negative
May 9, 2011	B1, negative watch
March 7, 2011	B1, negative
December 16, 2010	Ba1, negative watch
June 14, 2010	Ba1, stable
April 22, 2010	A3, negative watch
April 22, 2010	A3, negative
December 22, 2009	A2, negative
October 29, 2009	A1, negative watch
February 25, 2009	A1, stable
January 11, 2007	A1, positive
November 4, 2002	A1, stable

Source Trading Economics (2024)

10.3 Conclusion

This chapter was about digging into the third systemic bottleneck created through debt-based money, monetary hunger. This artificial force born out of the very nature of debt, public and private, exists in all debt-based economies, irrespective of capital accumulation. A growing economy does not eliminate this hunger, on the contrary. Money supply growth is achieved through public and private debts, and controlled, maintained, and monitored through default regulations and credit ratings. The artificial hunger is kept in line with a real whip.

Given a spaceless financial value framework, financial mathematics, and monetary architecture, the limitations imposed by debt-based money constrain our ability to create and invest the resources we need to address our expansion in outer space. Given the primacy they are given over our ecosystem, debt repayments are mandatory while emissions and space impact reporting are not, everyone would serve their debts before the environment, before space. As such, our ability to invest and create a sustainable reality on Earth is also undermined by these limitations.

Whether we are discussing our future in outer space or the energy transition aimed at addressing climate change, our spaceless value framework, mathematics and monetary architecture are primary evolutionary bottlenecks.

References

Bank of Greece. 2024. Government Benchmark Bond Prices and Yields. Bank of Greece. https://www.bankofgreece.gr/en/statistics/financial-markets-and-interest-rates/greek-government-securities. Accessed 26 June 2024.

CFA Institute. 2-024. Credit Default Swaps. The CFA Institute. https://www.cfainstitute.org/en/membership/professional-development/refresher-readings/credit-default-swaps. Accessed 12 June 2024.

ECB. 2012. Eligibility of Greek Bonds Used as Collateral in Eurosystem Monetary Policy Operations. European Central Bank. https://www.ecb.europa.eu/press/pr/date/2012/html/pr120228.en.html. Accessed 12 June 2024.

ECB. 2011. Financial Integration in Europe. European Central Bank. https://www.ecb.europa.eu/pub/pdf/fie/financialintegrationineurope201105en.pdf. Accessed 12 March 2024.

Gibson, H.D., Hall, S.G., Tavlas, G.S. 2024. The Greek Financial Crisis: Growing Imbalances and Sovereign Spreads. Bank of Greece Working Paper. 124, March 2024.

IIF. 2024. Global Debt Monitor: Navigating the New Normal. Institute for International Finance. https://www.iif.com/Products/Global-Debt-Monitor. Accessed 01 June 2024.

Investing. 2024. GRGV5YUSAC=R Overview. Investing.Com. https://www.investing.com/rates-bonds/greece-cds-5-years-usd-historical-data. Accessed 26 June 2024.

IMF. 2023. 2023 Global Debt Monitor. International Monetary Fund. https://www.imf.org/-/media/Files/Conferences/2023/2023-09-2023-global-debt-monitor.ashx. Accessed 12 April 2024.

Moody's. 2012. Moody's Comments on Greek Debt Exchange. https://www.moodys.com/research/Moodys-comments-on-Greek-debt-exchange-Announcement--PR_240125. Accessed 12 May 2024.

Papazian, A. 2022. *The Space Value of Money: Rethinking Finance Beyond Risk and Time*. New York: Palgrave Macmillan. https://doi.org/10.1057/978-1-137-59489-1.

Papazian, A. 2023. *Hardwiring Sustainability into Financial Mathematics: Implications for Money Mechanics*. New York: Palgrave Macmillan. https://doi.org/10.1007/978-3-031-45689-3.

Statista. 2023. Total Outstanding Public and Private Debt Across All Sectors in the United States from 2000 to 2022. https://www.statista.com/statistics/1083150/total-us-debt-across-all-sectors/. Accessed 2 February 2024.

Part IV

Breaking Risk and Time

Part II and III discussed our current financial value framework, financial mathematics, and monetary architecture, and revealed fundamental shortcomings that thwart and hinder our ability to invest and build sustainably on Earth and our ability to invest and expand in outer space.

This part of the book introduces the key transformations necessary to break the chains of risk and time and reconfigure our financial value framework, mathematics, and monetary architecture. It presents a space-adjusted framework and equations that can ensure both the sustainability and future expansion of human productivity.

11

Introducing Space

Going to the moon, seeing the moon up close was marvelous. But compared to the view of the Earth, I thought the moon was nothing. I thought that tiny little thing out my other window was the whole show. The Earth was just gorgeous. Tiny little shiny blue and white, a smear of rust across it that we called the continents, a glorious sight.
Michael Collins, Gemini 10 and Apollo 11 Astronaut, 1969

It is also conceivable that the flourishing of new global space systems may alter important cultural certainties of the past because the global reach of the sea-space continuum allows the (re)drawing of charts and routes that ignore traditional political and geographical borders. The scientific reading of the history of space and the understanding of the laws of physics and of life itself contribute to fostering the hope of a globally shared knowledge, that projects itself, with due respect, as an answer to fundamental questions about the existence and destiny of mankind
Franco Malerba, STS-46 Atlantis Astronaut, 1992

Space and outer space have been entirely absent from our financial value framework, mathematics, and debt-based monetary architecture. Our value principles and equations discriminate against our evolutionary investments. The three systemic bottlenecks created through our debt-based monetary architecture, i.e., calendar time, monetary gravity, and monetary hunger, inhibit our ability to create and invest all the required resources necessary for building a fair and sustainable reality on Earth as well as safe and functioning habitats in outer space.

A. V. Papazian, *Financing the Race to Space*, https://doi.org/10.1007/978-3-031-73102-0_11

This and the following chapters aim to introduce the necessary theoretical, conceptual, and mathematical adjustments that can address these evolutionary impediments. The first step is to introduce space, as an analytical dimension and our physical context, into the analytical framework of finance. Indeed, any civilisation or species in space that would like to live and expand in outer space, while using money and monetary incentives to drive and guide its own creativity and productivity, must surely integrate the dimension of space into its financial value framework, mathematics, and monetary architecture.

This chapter aims to do just that, to introduce space into finance. In other words, calendar time is no longer the only pillar of our analytical framework, and risk is not the only variable that matters. The introduction of the analytical dimension of space into finance implies that cash flows must also be assessed vis-à-vis space. To introduce space into finance, we need a proper conceptualisation of the analytical dimension that represents our physical context. Just like time is calendar time, and risk is the probability of a loss, space needs its own rational representation.

11.1 Space

As defined in chapter three, space is our physical context of matter, irrespective of constitution, composition, density, dynamics, and temperature, stretching from subatomic to interstellar space and every layer in between and beyond, where outer space is but a segment. Figure 11.1 is identical to the one introduced in chapter three and depicts a layered conceptualisation of our physical context.

As discussed previously, the layers depicted in Fig. 11.1 can be broken down into sublayers, and sublayers can be further defined when needed and relevant. Table 11.1 depicts examples of sublayers in the hydrosphere and the continental crust. When assessing our activities in space, sublayers may be critical and can have very specific contextual relevance. The same applies to all other layers of course.

Moreover, as discussed in chapter three, this conceptualisation of our physical context, of space, is a conceptualisation and it is for the benefit of the analytical solutions discussed in the following chapters. As such, like other conceptual lines, i.e., country borders, the Kármán line, and the prime meridian, it is designed to help us structure and organise ourselves and our productive activities on planet Earth and beyond.

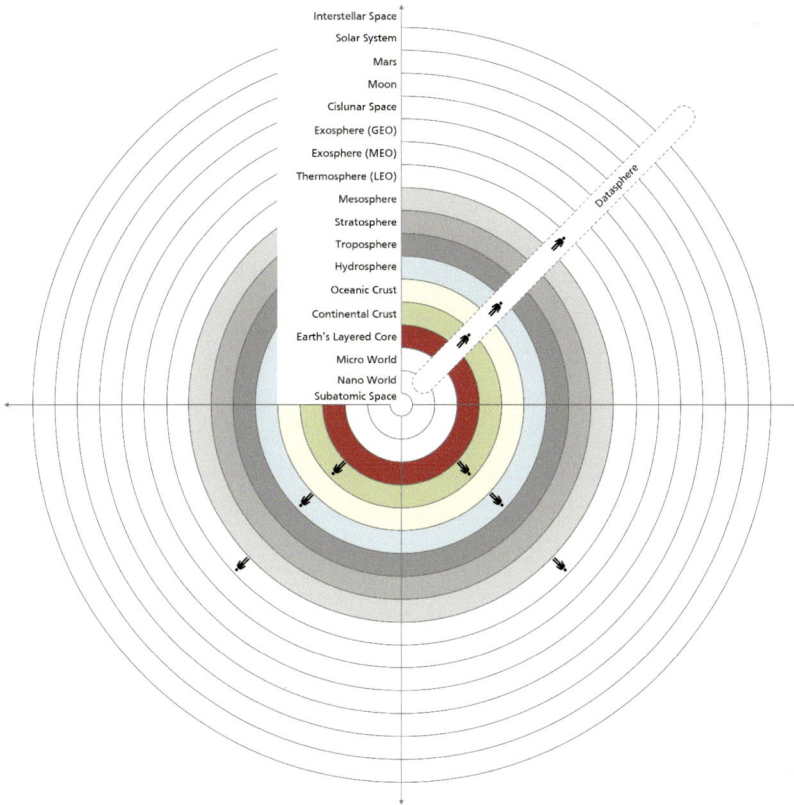

Fig. 11.1 Space layers and outer space (*Source* Adapted and updated from Papazian [2022])

11.2 Why Space Layers

The layered conceptualisation of space is important for the solutions I present in later chapters, and it is based on a number of factual observations. Our context of matter is made up of very different environments, which differ from each other in terms of constitution, composition, density, dynamics, and temperature. As such, the logic behind a layered conceptualisation of space is born out of the necessity to recognise that our activities have a very unique impact on different layers of space.

Moreover, some layers are more easily accessible than others. The hydrosphere, made up of our oceans, rivers, ice sheets, and lakes, is far more accessible to us than the mantle and inner core of our planet. Similarly, we breathe the air in our atmosphere by default, we have immediate access to the troposphere, but other layers, such as the stratosphere and exosphere, require

Table 11.1 Space sublayers: hydrosphere and continental crust

Space layers	Sub-layers	Sub-Layer Type Examples
Hydrosphere	Seas	
	Lakes -------->	Tectonic lakes
		Volcanic lakes
		Glacial lakes
		Fluvial lakes
		Solution lakes
		Landslide lakes
		Aeolian lakes
		Shoreline lakes
		Organic lakes
		Anthropogenic
		Meteorite lakes
	Rivers	
	Ice Sheets	
	Oceans	
	Epipelagic Zone - The Sunlight Zone	
	Mesopelagic Zone - The Twilight Zone	
	Bathypelagic Zone - The Midnight Zone	
	Abyssopelagic Zone - The Abyss	
	Hadal Zone - The Trenches	
Continental crust	Land Surface	Tundra
	Mountains	Taiga
	Built Up	Temperate broadleaf and mixed forest
		Temperate steppe
		Subtropical moist forest
		Mediterranean vegetation
	Vegetation -------->	Tropical forests
		Arid desert
		Xeric shrubland
	Cropland	Dry steppe
	Soil	Semiarid desert
	O Horizon - Organic Layer	Grass savanna
	A Horizon - Top Soil Nutrient Layer	Tree savanna
	E Horizon - Eluviation Layer	Tropical dry
	B Horizon - Subsoil Mineral Layer	Subtropical dry
	C Horizon - Regolith Layer	Tropical rainforest
	R Horizon - Bedrock Layer	Alpine tundra
		Montane forest
	Deep Crust	

Source Adapted from Papazian (2022)

special technologies. Some layers are relevant for our development, others not as much.

This layered conceptualisation is important from an economic and financial perspective. I identify five key reasons why this is so based on three aspects: access, activity, and cleanup.

Value Chain: Human productive and non-productive activities affect different layers of space. Even when in the same industry, the value chains of investments can affect different layers. For example, a shipping company that uses aeroplanes to transport its cargo affects the stratosphere, while a shipping company using ships affects the hydrosphere. Naturally, a shipping company using rockets to transport cargo to the ISS is affecting the Exosphere and all the layers of the atmosphere on the way.

Multi-Layer Footprint: Very often, the value chains and space footprint of our businesses and activities affect more than one space layer. Our investments can involve a different combination of space layers. The above-discussed shipping companies have their corporate headquarters somewhere on land, which means that they affect the continental crust as well. If they use digital technologies, and AI-managed platforms, they are also affecting the datasphere.

Impact Intensity: In parallel to the above, the impact of the same activity can differ across space layers. For example, Green House Gas (GHG) emissions in the stratosphere have a different impact from GHG emissions in busy densely populated cities with high rise buildings. The impact on humans and the surrounding environment is different, and thus, the same activity can have a very different impact across different layers of space.

Technology: Our activities involve and imply the necessity for different kinds of technologies. Some layers of space, such as the exosphere, require extensive technological intermediation. Meanwhile, other layers, like the troposphere, do not. This is also relevant for activity and cleanup. Travelling across the continental crust can involve different kinds of technological support, but travelling in cislunar space is an entirely different challenge. Similarly, cleaning waste or debris from our oceans, from our rivers, from our streets, from our food chain, and from Earth orbit requires very different technologies.

Costs: Last but not least, given all of the above, the costs of space impact differ across different layers of space, whether it is for access, activity, or cleanup.

11.3   Space Layers in Metres

The conceptualisation of space presented in Fig. 11.1 defines space as our physical context of matter stretching from subatomic to interstellar space and every layer in between and beyond. Introducing this context into finance from an analytical perspective can also be presented in pure measurement form. A more abstract conceptualisation, see Fig. 11.2, can be depicted through a three-dimensional space of length l, starting from $l = 10^0$ m (1 m) and stretching towards the astronomical and the infinitesimal (Papazian 2023, 72). Figure 11.2 depicts layers from 10^{20} to 10^{-20} for convenience.

In truth, despite the limits of our own observable universe, there is no outer boundary to a three-dimensional measurement of space. However, when looking within matter, our current understanding reveals that at the

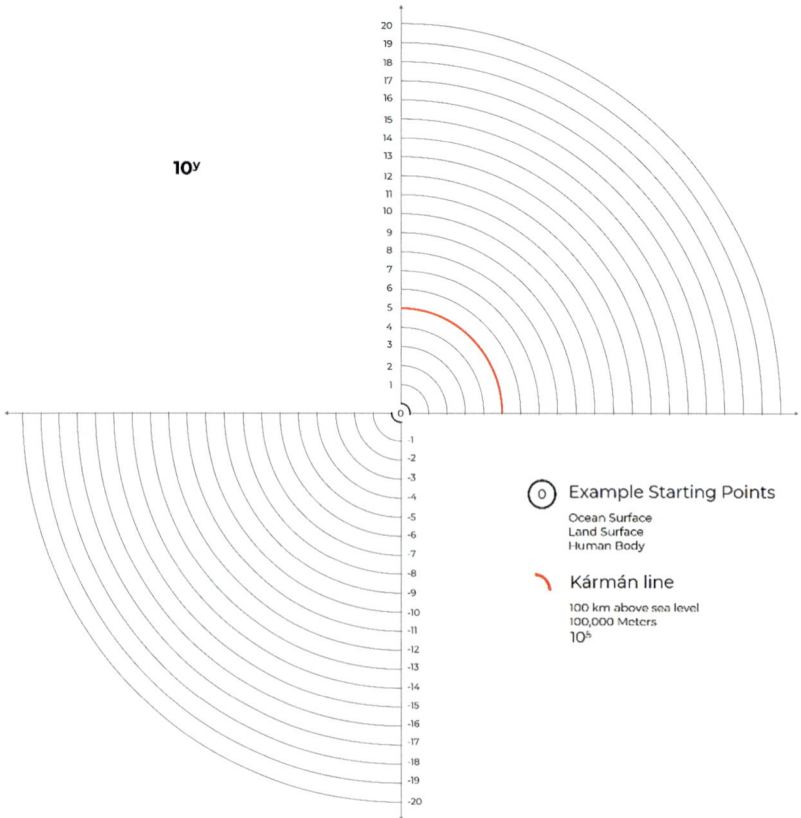

Fig. 11.2 Space layers in metres (10^y) from any point of matter (*Source* Adapted and updated from Papazian [2023])

scale of the Planck length = 1.616255×10^{-35} m, our conventional under-standings of space–time cease to be relevant (Padmanabhan 1985; Seiberg 2006; Hossenfelder 2012). Table 11.1 provides random examples of 'things' in different layers (Table 11.2).

It is critically important to realise that an abstract measurement-based conceptualisation of space and its many layers can be described from any point of matter, anywhere in the universe. Indeed, as noted in Fig. 11.2, example starting points relevant to our discussion can be the human body, land surfaces, and ocean surfaces. In Fig. 11.2, the Kármán line, 100 km above sea level, is identified in red, at 10^5.

I have suggested previously that a thorough investigation and mapping of our space impact may involve the assessment of our activities based on both, Figs. 11.1 and 11.2. This is so because our starting point, like the human body, ocean surface, or land surfaces, can define a very different composition and context within matter.

Table 11.2 Space layers by length of 3D cube

Y axis in metres	Equivalent	Random example in range—diverse online sources
10^{30}	1 quettametre	
10^{29}	100 ronnametre	New Prefix adopted in November 2022 by the CGPM (2022)
10^{28}	10 ronnametre	
10^{27}	1 ronnametre	
10^{26}	100 yottametres	The radius of the observable universe from Earth is approximately 435.19 yottametres
10^{25}	10 yottametres	GN-Z11 the high-redshift galaxy is 302.74 yottametres from Earth
10^{24}	1 yottametre	The distance to the Shapley Supercluster of galaxies is approximately 6.1 yottametres
10^{23}	100 zettametres	The Messier 87 supergiant elliptical galaxy is 501.41 Zettametres from Earth
10^{22}	10 zettametres	The Andromeda galaxy is approximately 23.7 zettametres away from Earth
10^{21}	1 zettametre	The Milky Way galaxy is approximately 1 zettametre across
10^{20}	100 exametres	The cluster of young stars 'Westerlund 2' is approximately 189.21 exametres from Earth
10^{19}	10 exametres	The exoplanet Kepler-443b is approximately 28.38 exametres away from Earth
10^{18}	1 exametre	Our sun's closest twin star HIP 56,948 is approximately 1.96 exametres away from Earth
10^{17}	100 petametres	The radius of the radio bubble emitted from Earth is approximately 946.073 petametres
10^{16}	10 petametres	The Alpha Centauri triple star system is 41.32 petametres from Earth
10^{15}	1 petametre	Light travels 9.461 petametres in one year
10^{14}	100 terametres	Light travels 777.062 terametres in one month
10^{13}	10 terametres	Voyager 1 was 23.953 terametres from Earth by 28/07/23, most distant human made object
10^{12}	1 terametre	Uranus is approximately 2.72395 terametres away from Earth
10^{11}	100 gigametres	Our sun is approximately 149.6 gigametres away from Earth
10^{10}	10 gigametres	Venus is approximately 41.4 gigametres from Earth
10^{9}	1 gigametre	Mars is approximately 7.834 gigametres away from Earth

Y axis in metres	Equivalent	Random example in range—diverse online sources
10^8	100 megametres	The Moon is 384.399 megametres away from Earth
10^7	10 megametres	The Galileo Satellites orbit the Earth at an altitude of 23.222 megametres
10^6	1 megametre	The Cluster C1 satellite is on orbit at an altitude of 9 megametres
10^5	100 kilometres	The International Space Station (ISS) orbits the Earth at ≈402.336 kilometres from sea level
10^4	10 kilometres	Commercial airlines fly their aeroplanes at an average altitude of 10.6 kilometres
10^3	1 kilometre	Everest's summit is at 8.849 kilometres from sea level
10^2	1 hectometre	The height of the Empire State Building to its tip is 4.43 hectometres
10^1	1 decametre	The length of a wind turbine blade ranges between 5.2 and 10.7 decametres
10^0	1 metre	The average height of humans is 1.7 metres
10^{-1}	1 decimetre	The average length/height of a full-term newborn human baby is 5 decimetres
10^{-2}	1 centimetre	A 2litre soda plastic bottle is 30 centimetres long
10^{-3}	1 millimetre	The average length of an adult mosquito is 4.5 millimetres
10^{-4}	100 micrometres	The average thickness of an eggshell is 300 micrometres or microns
10^{-5}	10 micrometres	The average thickness of a human hair is 70 micrometres or microns
10^{-6}	1 micrometre	The average thickness of a human red blood cell is 6–8 micrometres or microns
10^{-7}	100 nanometres	The diameter of the Corona virus (SARS-CoV-2) ranges between 50 and 140 nanometres
10^{-8}	10 nanometres	The diameter of smallest viruses like Adeno-Associated Virus (AAV) is 20 nanometres
10^{-9}	1 nanometre	The diameter of a strand of human DNA is 2.5 nanometres
10^{-10}	100 picometres	The average size of a water molecule is 280 picometres
10^{-11}	10 picometres	The Bohr radius of a hydrogen atom is approximately 53 picometres
10^{-12}	1 picometre	The Compton wavelength of an electron is 2.4263 picometres
10^{-13}	100 femtometre	The diameter of the atomic nucleus of Uranium is approximately 11.7 femtometres or fermi

(continued)

Table 11.2 (continued)

Y axis in metres	Equivalent	Random example in range—diverse online sources
10^{-14}	10 femtometre	The radius of a gold nucleus is approximately 8.45 femtometres or fermi
10^{-15}	1 femtometre	The classical radius of an electron is 2.81 femtometres or fermi
10^{-16}	100 attometres	The approximate radius of a Proton is 841.8 attometres
10^{-17}	10 attometres	The range of the weak nuclear force is estimated to be 10 attometres
10^{-18}	1 attometre	The upper limit of the diameter of quarks which make up protons and neutrons in an atom
10^{-19}	100 zeptometres	
10^{-20}	10 zeptometres	Too small to give any non-technical examples
10^{-21}	1 zeptometres	
10^{-22}	100 yoctometre	
10^{-23}	10 yoctometre	
10^{-24}	1 yoctometre	
10^{-25}	100 rontometre	New Prefix adopted in November 2022 by the CGPM (2022)
10^{-26}	10 rontometre	
10^{-27}	1 rontometre	
10^{-28}	100 quectometre	
10^{-29}	10 quectometre	
10^{-30}	1 quectometre	

Source Papazian (2023) inspired by Eames and Eames (1977)

11.4 Conclusion

Risk and Time cannot possibly be the only two analytical axes of our financial value framework. Decades of intellectual history, dozens of Nobel Prizes, and many millions of academic papers in finance are built entirely on risk and time, without any reference to space, our physical context of matter, and our impact on it.

I offer a layered conceptualisation of space as it is an authentic representation of our context and the many environments within it. It provides us the ability to mirror our context with all its wealth and complexity in a simple and useful format.

As you can see from the above discussion, the introduction of space into finance, as analytical dimension and our physical context, goes far beyond using geospatial data to analyse investments (CGFI-SFY 2021). It also goes far beyond measuring trade and currency exchange dynamics between different civilisations in outer space (Krugman 1978; Gaarder Haug 2004).

Indeed, the introduction of space into finance as discussed here is a fundamental transformation of our analytical construct upon which our framework is based. Space must now be formally integrated into our financial value framework, mathematics, and monetary architecture—where space is our physical context of matter, irrespective of constitution, composition, density, dynamics, and temperature, stretching from subatomic to interstellar space and every layer in between and beyond, including outer space.

In other words, cash flows must now be assessed vis-à-vis space as well, and not just in relation to risk and time.

References

CGFI-SFI. 2021. State and Trends in Spatial Finance. Centre for Green Finance and Investment—Spatial Finance Initiative. https://www.cgfi.ac.uk/wp-content/uploads/2021/07/SpatialFinance_Report.pdf. Accessed 2 February 2022.

Eames, C., Eames, R. 1977. Powers of Ten. Eames Office. http://www.powersof10.com/. Accessed 14 June 2023.

Gaarder Haug, E., 2004. Space-Time Finance: The relativity Theory's Implications for Mathematical Finance. *Wilmott Magazine*, July, 2–15.

Hossenfelder, S. 2012. Can We Measure Structures to a Precision Better Than the Planck Length? Classical and Quantum Gravity 29 (11). https://doi.org/10.1088/0264-9381/29/11/115011. Accessed 2 August 2023.

Krugman, P. 1978. The Theory of Interstellar Trade. *Economic Enquiry*, 48, 1119–1123.

Padmanabhan, T. 1985. Physical Significance of Planck Length. *Annals of Physics* 165: 38–58. https://doi.org/10.1016/S0003-4916(85)80004-X. Accessed 30 July 2023.

Papazian, A. 2022. *The Space Value of Money: Rethinking Finance Beyond Risk and Time*. New York: Palgrave Macmillan. https://doi.org/10.1057/978-1-137-594 89-1.

Papazian, A. 2023. *Hardwiring Sustainability into Financial Mathematics: Implications for Money Mechanics*. New York: Palgrave Macmillan. https://doi.org/10.1007/978-3-031-45689-3.

Seiberg, N. 2006. *Emergent Spacetime*. School of Natural Sciences, Institute for Advanced Study. Princeton. https://arxiv.org/pdf/hep-th/0601234.pdf. Accessed 30 March 2023.

12

Respecting Space

The Earth was small, light blue, and so touchingly alone, our home that must be defended like a holy relic.
Alexey Leonov, Voskhod 2 Cosmonaut, 1965

What still struck me as impressive was the shining blue Earth, which looked like one form of life floating in the universe. At the same time, I was reminded of the thinness of the blue layer, which is the atmosphere… Such a thin atmosphere protects every living thing – forests, trees, fish, birds, insects, human beings and everything.
Toyohiro Akiyama, Soyuz TM-11 TBS Correspondent, 1990

Having conceptualised and introduced space, our physical context of matter that stretches from subatomic space to interstellar space and every layer in between and beyond, we now have a functional definition that can be used as an analytical dimension in finance. To assess cash flows vis-à-vis space, we must define our relationship with this newly introduced dimension.

As discussed in Chapter 5, our current financial value framework is built around two core principles of value, Risk and Return and Time Value of Money, which define our relationship with risk and time respectively. These principles are used and applied in our core finance equations to assess the value of cash flows. Moreover, they serve only one stakeholder, the mortal risk-averse return-maximising investor.

Thus, a space-adjusted financial value framework requires a new principle of value that can complement the existing two and can define the logic through which we assess the *space value of cash flows*. I have proposed the

A. V. Papazian, *Financing the Race to Space*, https://doi.org/10.1007/978-3-031-73102-0_12

introduction of a new principle into our core value framework, the Space Value of Money. In other words, financial analysis must take into account the risk, time, and space value of money when assessing cash flows, profits, and returns.

This is the first step towards the solutions that will allow us to address the challenges presented in previous chapters. It is the principle that defines our relationship with space, our physical context, and establishes respect for space.

12.1   The Space Value of Money

The Space Value of Money (SVoM) principle defines our relationship with space and provides the rationale through which we can assess the space value of cash flows. Based on the evidence presented in Chapter 4, our treatment of and impact on space leaves a lot to be desired. Indeed, we have littered every space layer we have come in contact with, and we have done so without any consideration of the damages we inflict on ourselves and future generations. Thus, the space value of money introduces that which we need the most, respect for space and its many layers. Respect for space is the first building block of a much larger framework.

> The space value of money principle complements time value of money and risk and return. It establishes our spatial responsibility and requires that a dollar ($1) invested in space has at the very least a dollar's ($1) worth of positive impact on space. (Papazian 2022, 104)

The space value of money is a necessary first step to trigger the necessary transformations in our value framework. The core function of the principle is to establish and entrench our responsibility for space impact, and to establish the bottom threshold of investment acceptability. In other words, a space-adjusted financial value framework integrates the respect for our physical context as a primary requirement alongside the minimisation of risks and maximisation of returns.

By requiring that a dollar invested in space has at the very least a dollar's worth of positive impact on space, taking into account all layers, the space value of money principle entrenches sustainability into our productive capacity, as it conditions capital allocation and money creation by a positive space impact principle, and establishes foundations upon which space impact is valued and integrated into our models. This is critically important.

The issue we face with our current financial value framework is not just the absence of space, but also the absence of any consideration of our space impact. To date, space impact is not included in our equations of value and return in finance. In other words, positive or negative, space impact is not considered relevant to the equations that aim to derive the value of cash flows.

The introduction of the space value of money principle, therefore, alongside responsibility, introduces space impact as an integral element of value. This is particularly relevant to outer space investments. As the reader will recall, our risk and time dominated framework is built on two principles that discriminate against our evolutionary investments with distant cash flows and high risks. Two features that characterise all outer space investments. The third feature that is abundant in outer space exploration and settlement projects is space impact. Indeed, the expansion and extension of our footprint beyond the atmosphere has an intrinsic evolutionary value. Through the space value of money principle, our space impact becomes an integral part of our equations.

Indeed, the space value of money principle establishes our responsibility vis-à-vis space and makes the space impact of investments a central element of our value framework. As I will discuss and expand in later chapters, this addresses the challenges of an entirely risk and time-based construct and opens the door to an analytical framework that does not discriminate against projects that have high risks and distant returns, i.e., all our evolutionary investments.

Figure 12.1 and Fig. 12.2 depict what I have called the Transition Return Impact Map (The TRIM), which helps us visualise the functional relevance of this new principle. The space value of money allows the allocation of capital to investments that have a positive space impact, while allowing the integration of such impact into the value equations of finance.

This is necessary because our current value framework and equations do not prevent investors from investing in opportunities that have a negative space impact, left quadrant in Fig. 12.1 (Quadrant 3). The current state of the world and our impact on space is evidence of this unwritten permission. Given our current framework and equations in finance, the bottom two quadrants in Fig. 12.1 and Fig. 12.2 (Quadrants 1 and 2) are unattractive because expected returns are negative.[1]

---

[1] Note that actual returns could be negative post investment, and this is referring to expected and required returns. Also, some public investors may initiate investments in the bottom two quadrants for a variety of reasons, amongst them is the provision of public goods, paid through tax revenue.

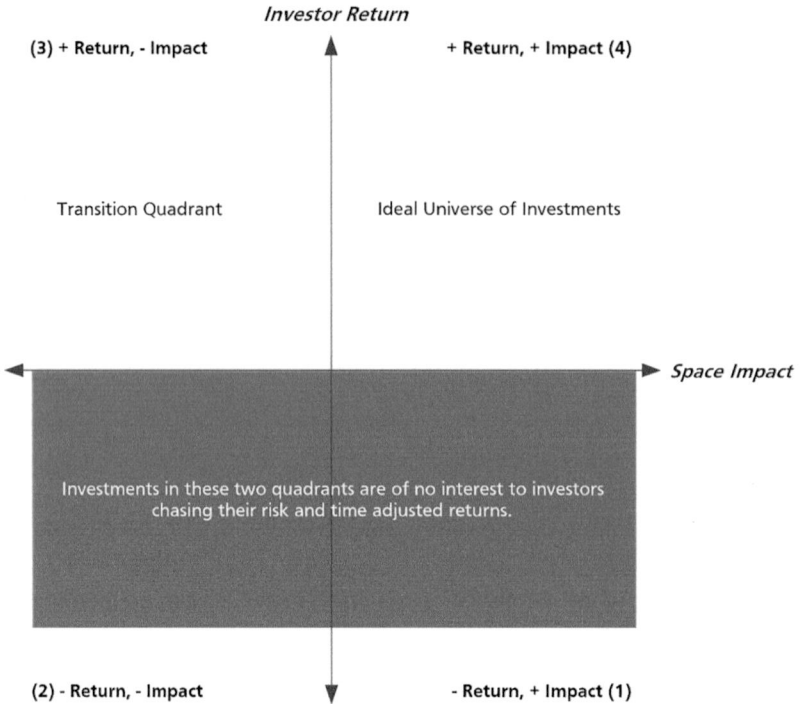

Fig. 12.1 The TRIM: Transition Return Impact Map (*Source* Papazian [2023])

The space value of money principle allows us to prevent new investments that have a negative impact on space, which allows us to transform our impact on space going forward.

Without the space value of money principle, the situation we are faced with now, our financial value framework is blind to space and the space impact of investments and does not consider space impact as relevant. This is the theoretical junction where finance theory absolves investors of their responsibility for ecological destruction.

12.2 The New Stakeholders of Finance

The introduction of the space value of money principle has fundamental implications that help us define the new space-adjusted equations. It also introduces planet and humanity as equal stakeholders of our financial value framework. The mortal risk-averse return-maximising investor is not the one and only stakeholder of our framework and mathematics, not anymore. A

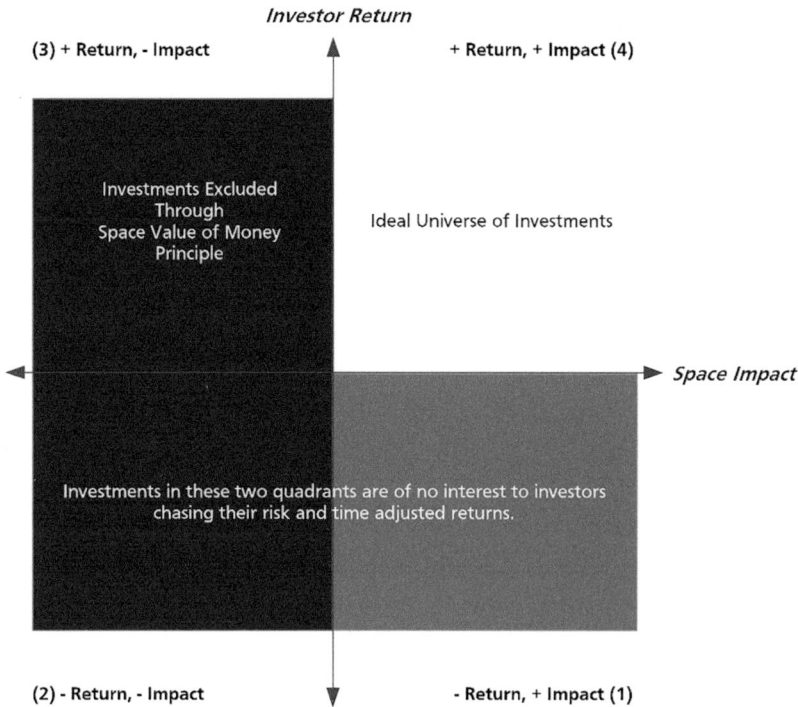

Fig. 12.2 The TRIM with Space Value of Money (*Source* Adapted from Papazian [2023])

pollution-averse planet and an aspirational human society are now equal stakeholders.

In Papazian (2022) I introduce a pollution-averse planet and an aspirational human society as the two new stakeholders of finance which must be on an equal footing with the mortal risk-averse return-maximising investor. This is fundamental for any meaningful and effective change.

This may sound similar to the popular and commonly used expression 'people and planet before profit,' but it is not really the same. The principle establishes planet and humanity as equal stakeholders without downplaying the priorities of the investor. It simply introduces a better and more balanced equation. The rational behind this is simple. The respect for the collective and the planet cannot trample upon the individual in the chain. The parts are the whole, even if the whole may have different priorities. As discussed earlier, our challenge is to find a balanced framework that recognises the matrix of priorities that links an eternal whole (humanity through procreation) to its mortal parts (individual humans).

This distinction is important. The space value of money principle does not aim to be prohibitive and or constraining to private enterprise. It may be considered so at the moment given our current monetary architecture and the state of the world, but it is not so in essence. The purpose is to provide a tool through which investors can better define the universe of acceptable investments. Moreover, as I discuss in later chapters, the space value of money applies to individuals and businesses but also money creators.

12.3   Conclusion

Introducing space into finance is the first step. The next step is to introduce the principle that defines our relationship with this new dimension. Given the discussion up to this point, it is explicitly clear that what we are missing is respect for space. The space value of money principle establishes and entrenches that respect into our value framework.

"The space value of money principle complements time value of money and risk and return. It establishes our spatial responsibility and requires that a dollar ($1) invested in space has at the very least a dollar's ($1) worth of positive impact on space" (Papazian 2022, 104). Through the space value of money principle we can balance our framework and address another important omission. From now on, space impact is an integral element of value.

This is critical if we are to rectify the biases of our current framework and correct the mispricing of our evolutionary investments. The risk and return and time value of money principles do not prevent negative impact on their own, and they misprice highly risky and very distant returns—two prominent features of our evolutionary investments. Indeed, outer space exploration and settlement projects involve high risks, distant returns, and abundant space impact.

The space value of money principle is the starting point of a larger framework that can remove the theoretical and structural obstacles that undermine our ability to create a sustainable and fair reality on Earth and frustrate our ambitions to expand in outer space.

References

Papazian, A. 2023. *Hardwiring Sustainability into Financial Mathematics: Implications for Money Mechanics*. New York: Palgrave Macmillan. https://doi.org/10.1007/978-3-031-45689-3.

Papazian, A. 2022. *The Space Value of Money: Rethinking Finance Beyond Risk and Time*. New York: Palgrave Macmillan. https://doi.org/10.1057/978-1-137-59489-1.

13

Valuing Space

Once you've been in space, you appreciate how small and fragile the Earth is.
Valentina Tereshkova, Vostok 6 Cosmonaut, 1963

You are mesmerised looking out the window of your spacecraft. As there's no work to do in the capsule in the next hours and you don't feel like sleeping, you observe the show of the Earth and the sky. The spacecraft turns and turns around its axis and displays to you one landscape after another.
Pedro Duque, STS-95 Discovery Astronaut, 1998

Now that we have introduced space into finance and we have clarified the principle that defines our relationship with space, we must consider the implications of these changes on our financial mathematics. In other words, once space, as analytical dimension, and our physical context, is introduced into our financial value framework, and our responsibility for space impact is established through the space value of money principle, we can begin to design the space-adjusted equations that integrate the space impact of cash flows and investments.

Let me state from the outset that you do not need to be a mathematician or a quant to understand these equations. They are simple calculations, and their meaning and essence are explained in the text in plain English.

© The Author(s), under exclusive license to Springer Nature Switzerland AG 2024
A. V. Papazian, *Financing the Race to Space*,
https://doi.org/10.1007/978-3-031-73102-0_13

13.1   Space-Adjusted Financial Mathematics

13.1.1   The Double Timeline

Fig. 13.1 is the standard cash flow timeline we first discussed in chapter six. It is very often used in finance textbooks to describe the future cash flows that must be discounted to the present when measuring the risk and time value of cash flows. As demonstrated earlier, the current focus of our equations in finance is the future expected cash flows, and the space impact it would take to achieve or expect them is not part of the equations. The mathematical junction where responsibility and impact are abstracted.

Given the introduction of space and our responsibility for impact in the previous chapters, we are now faced with the need to integrate the space impact of investments and cash flows into their valuation.

I have introduced the double timeline as a first step to help integrate the space impact of cash flows into the analysis. Figure 13.2, at the most abstract level, allows us to visualise the cash flow timeline by taking into account space. As you can see, it is now divided into two: the risk timeline and space timeline. The ultimate purpose here is to account for the space impact of cash flows, across all affected space layers, in parallel to the cash flow expectations of the investor.

While a very simple change, this implies a profound transformation in what matters and what must be considered in our equations. Financial analysis can no more be focused on the future expected cash flows alone, prioritising the perspective of the mortal risk-averse return-maximising investor. With space and the space value of money principle, planet and humanity are equal stakeholders.

The double timeline recognises this change and facilitates the integration of the space impact of investments, across all the impacted layers. In other words, the space impact it takes to expect and/or achieve the future cash

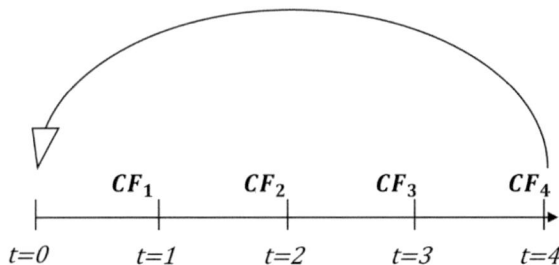

Fig. 13.1   The cash flow timeline (*Source* Author)

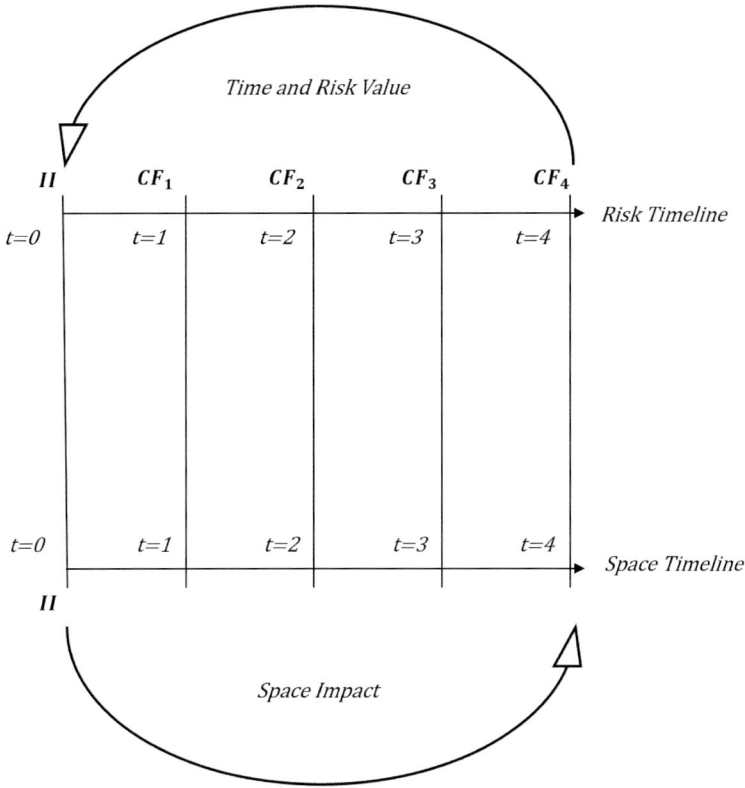

Fig. 13.2 The double timeline (*Source* Papazian 2022)

flows is now integral to the valuation exercise. Figure 13.3 is a variation of the double timeline that facilitates the mapping and quantification of an investment's impact across the affected layers. Ocean and Soil and their sublayers are given as an example of how this conceptualisation can be further refined depending on the value chain of the investment under consideration.

13.1.2 Gross and Net Space Value of Money

The minimum space value condition, given the space value of money principle, is that a dollar invested in space must, at the very least, have a dollar's worth of positive impact on space. I have introduced the concepts and equations of Gross Space Value and Net Space Value to measure the aggregate space impact of cash flows or investments (Papazian 2023, 2022).

$$\text{Gross Space Value}_{T,S} = \text{Net Space Value} + \text{Inital Investment} \qquad (13.1)$$

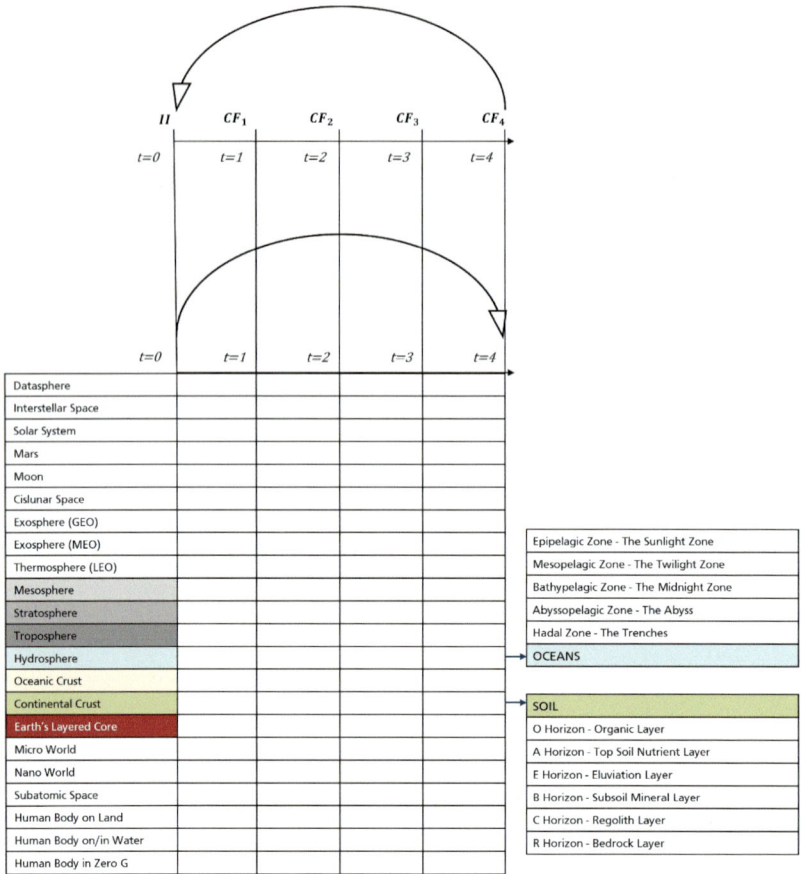

Fig. 13.3 The double timeline with space layers and example sublayers (*Source* Author)

$$NSV + II = GSV$$

When Net Space Value (NSV) is equal to zero, or the investment is *space neutral*, the space value of money condition is met. Gross Space Value is equal to the initial investment and thus a dollar invested in space meets the minimum requirement and has a dollar's worth of positive impact on space.

$$\mathbf{GSV = II}$$

The Net Space Value is defined as the Planetary, Human, and Economic impact of the investment or cashflows across all the layers of space affected and across all time periods. Each of the above elements, i.e., planetary, human, and economic, is further defined as the pollution, biodiversity, human capital, R&D, new money, and new asset impacts of the investment. Equation 13.2 is a conceptual summary of what the Net Space Value includes.

$$\mathbf{NSV}_{T\&S} = \textbf{Net Space Value of Investment}$$

$$\mathbf{NSV}_{T\&S} = \left\{ \begin{array}{l} \textbf{Planetary, Human,} \\ \textbf{and Economic Impact} \end{array} \right\}_{\textbf{All } S \textbf{ Layers \& } T \textbf{ Periods}} \tag{13.2}$$

T = Total Number of Years of Investment being Considered
S = All Space layers Involved in the Investment

$$\mathrm{NSV}_{T\&S} = \sum_{t=1}^{T}\sum_{s=1}^{S} \text{Pollution \& Biodiversity Impact}$$

$$+ \sum_{t=1}^{T}\sum_{s=1}^{S} \text{Human Capital \& R and D Impact}$$

$$+ \sum_{t=1}^{T}\sum_{s=1}^{S} \text{New Asset \& New Money Impact}$$

Table 13.1 lists the equations for each of the components of Net Space value. These equations measure the space impact of investments across *all the layers of space and across all time periods* (the two sigma $\sum$ operators). This is a unique contribution of the framework and allows a detailed accounting of the space impact of cash flows depending on the value chains involved. As discussed in chapter eleven, our productive and nonproductive activities affect space layers very differently, and these equations consider that diversity and account for it across all aspects, and at all time periods.

In the Net Space Value equation, at the very top of the table, the coefficient g (governance) is used as an aggregate measure of trust in the entity under consideration. "The coefficient g may or may not be considered relevant to a specific investment. When it is, it should be a coefficient between

Table 13.1 The equations of impact

Impact aspect	Net Space Value	$g \times (\mathbf{PI}_{T,S,P} + \mathbf{BI}_{T,S,B} + \mathbf{HCI}_{T,S} + \mathbf{RDI}_{T,S,N} + \mathbf{NAI}_{D,S,A} + \mathbf{NMI}_T)$	(13.3)
Planetary	Pollution Impact	$PI_{T,S,P} = \displaystyle\sum_{t=1}^{T}\sum_{s=1}^{S}\sum_{p=1}^{P} Q_{pst} \times C_{pst}$	(13.4)
	Biodiversity Impact	$BI_{T,S,B} = \displaystyle\sum_{t=1}^{T}\sum_{s=1}^{S}\sum_{b=1}^{B} Ab_{st} \times R_{bst}$	(13.5)
Human	Human Capital Impact	$HCI_{T,S} = f \times \displaystyle\sum_{t=1}^{T}\sum_{s=1}^{S} E_{st} + T_{st} + H_{st} + I_{st} + C_{st} + S_{st}$	(13.6)
	R and D Impact	$RDI_{T,S,N} = \displaystyle\sum_{t=1}^{T}\sum_{s=1}^{S}\sum_{n=1}^{N} h_n \times RD_{tsn}$	(13.7)
Economic	New Asset Impact	$NAI_{D,S,A} = \displaystyle\sum_{s=1}^{S}\sum_{a=1}^{A} k_a \times BVA_{asD}$	(13.8)
	New Money Impact	$NMI_T = (II \times DR \times BLR) + mm \times (II + X_T - M_T)$	(13.9)
Coefficients	Fairness	f	
	Health	h	
	Transition	k	
	Governance	g	

Source Papazian (2022)

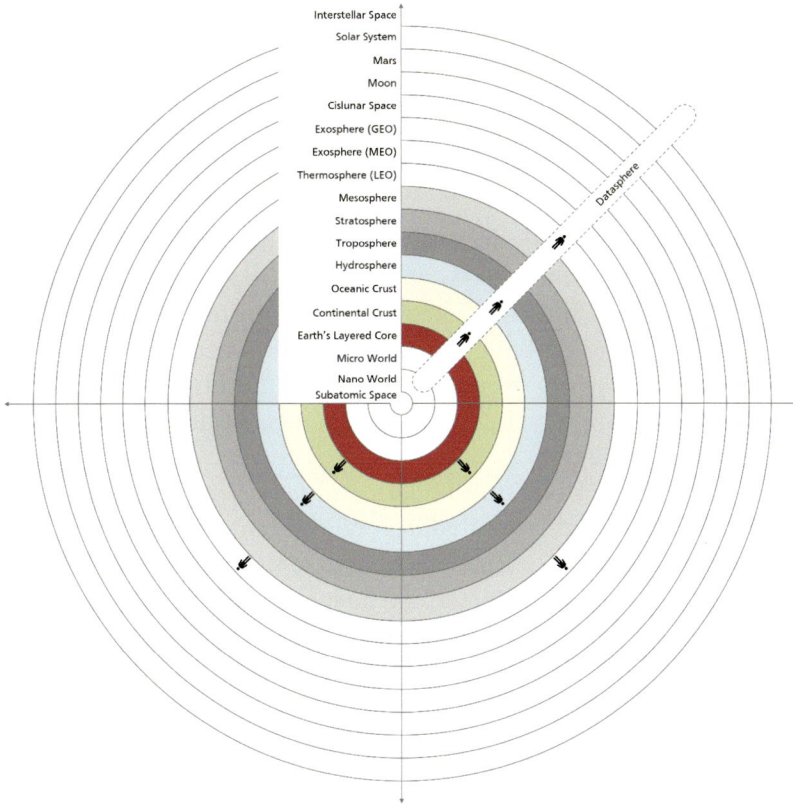

Fig. 13.4 Space layers (*Source* Author adapted and updated from Papazian [2022])

$0 \leq g \leq 1$ reflecting the level of trust in the veracity of the figures based on the corporate governance measures and structures in the investment. Transparency and accountability could be measured or quantified through the coefficient g through a number of different metrics" (Papazian 2022, 153).

In the following subsections, I expand on each of these equations without going into the detailed derivation process. You can find further details in Papazian (2023, 2022). My aim here is to explain their purpose and function.

In Fig. 13.4, I share the space layers diagram again as the equations of impact refer to them in the next subsections.

13.1.2.1 Pollution Impact

$$\mathrm{PI}_{T,S,P} = \sum_{t=1}^{T}\sum_{s=1}^{S}\sum_{p=1}^{P} Q_{pst} \times C_{pst} \qquad (13.10)$$

T = Total Number of Years in the Investment
P = Number/Types of Pollutants Involved
S = All Space Layers Involved
Q_{pst} = Quantity of Pollutant p in time t in space layer s
C_{pst} = Cost of Cleanup for Pollutant p in time t in space layer s

The pollution impact of the investment or cash flows considers all pollutants created/cleaned in all time periods and in all space layers and measures the clean-up cost through addition. The sign of the impact is positive when the investment is cleaning up pollution, and negative when it is causing pollution. The multiple sigma operators ($\sum$) serve the purpose of identifying and summing the impact of different pollutants, in different space layers, in different time periods.

This is an important aspect of impact, and measuring and accounting for pollution impact is necessary to address our current inability to invest and build sustainably here on Earth. Naturally, applying the same formula as we move out of Earth ensures that our financial mathematics accounts and prevents pollution before the fact, and not after the fact. This is critical to ensure that we do not deform the new planets we land on.

13.1.2.2 Biodiversity Impact

$$BI_{T,S,B} = \sum_{t=1}^{T} \sum_{s=1}^{S} \sum_{b=1}^{B} A_{bst} \times R_{bst} \qquad (13.11)$$

T = Total Number of Years in the Investment
S = All Space Layers Involved
B = All Types of Biodiversity Involved
$BI_{(T,S)}$ = Sum of All Biodiversity Impacts Across Years and Space Layers
A_{bst} = Area (ha) of Biodiversity Impact b in time t in space layer s
R_{bst} = Cost of Restoration $\left(\frac{\$}{ha}\right)$ of Biodiveristy Impact b in t in s

The biodiversity impact of the investment or cash flows considers all biodiversity loss/restoration created in all time periods and in all space layers and measures the restoration cost through addition. The sign of the impact is positive when the investment is restoring biodiversity, and negative when it is causing biodiversity loss. The multiple sigma operators ($\sum$) serve the purpose of identifying and summing different biodiversity impacts, in different space layers, in different time periods.

Measuring and accounting for biodiversity impact are necessary to address our current inability to invest and build sustainably here on Earth. Like in the case of pollution, applying the formula as we move out of Earth ensures that our financial mathematics accounts and prevents negative biodiversity impact before the fact, and not after the fact. This, again, is critical to ensure that we terraform appropriately, and we do not repeat the same mistakes we have witnessed here on Earth.

13.1.2.3 Human Capital Impact

$$\text{HCI}_{T,S} = f \times \sum_{t=1}^{T} \sum_{s=1}^{S} E_{st} + T_{st} + H_{st} + I_{st} + C_{st} + S_{st} \qquad (13.12)$$

S = All Space Layers Involved
T = Total Number of Years in the Investment
f = Coefficient of Fairness
E_{st} = Employment Expendiuture
T_{st} = Training and Education Expendiuture
H_{st} = Health Related Expendiuture
I_{st} = Immigration Related Expendiuture
C_{st} = Compensation Expenditure
S_{st} = Social Investment Expendiuture

The human capital impact of the investment or cash flows considers all expenditure and investment in humans within and outside the project, in all time periods, and in all space layers. The multiple sigma operators ($\sum$) serve the purpose of identifying and summing human impact in different space layers and in different time periods. This ensures that humans on the continental crust, on the hydrosphere, and in the thermosphere are always treated and supported properly.

Measuring and accounting for human capital impact is necessary to address our current inability to invest and build a fair and sustainable reality here on Earth, as well as beyond Earth. The coefficient f is a coefficient of fairness, as the absolute value of expenditure does not always reflect the fairness, balance, and just treatment of humans affected.

The coefficient f is a number between $-1 \leq f \geq 1$, that identifies the level of fairness in the organisation. Naturally, it can be considered a subjective measure. While this is true. It is necessary to ensure that investments abide by our standards of fairness and decency. Moreover, it allows us to assign a negative sign to human capital expenditure for a number of extreme cases of injustice and abuse.

13.1.2.4   R&D Impact

$$\text{RDI}_{T,S,N} = \sum_{t=1}^{T} \sum_{s=1}^{S} \sum_{n=1}^{N} h_n \times \text{RD}_{tsn} \tag{13.13}$$

h = Coefficient of health
$\text{RD}_{(T,S,N)}$ = R&D Expenditure per Project N across T, space layers S
S = All Space Layers Involved
T = Total Number of Years in the Investment
N = Number/All R and D Projects Involved in the Investment

The R&D impact of the investment or cash flows considers all expenditure into research and development projects within the investment. R&D impact is quantified on a project basis as investment opportunities may involve multiple parallel R&D projects. The formula measures R&D expenditure in all projects, time periods, and in all space layers. The multiple sigma operators ($\sum$) serve the purpose of identifying and summing the different projects, in different space layers and in different time periods.

Measuring and accounting for R&D impact is necessary to address the necessity to invest in new knowledge and technologies on our way to new habitats. The coefficient h_n is equal to either -1 or 1, to identify the healthy management of the R and D project. It must take into account the safety and protection of the environment as well as the individuals involved. The coefficient h_n as a coefficient of health is necessary because the absolute value of R&D expenditures does not always reflect the potential negative and/or toxic implications of a mismanaged process.

13.1.2.5   New Asset Impact

$$\text{NAI}_{D,S,A} = \sum_{s=1}^{S} \sum_{a=1}^{A} k_a \times \text{BVA}_{asD} \tag{13.14}$$

k_a = Coefficient of Transition Value
S = All Space Layers Involved
D = Period /Date When Asset is Created and Added in Books
A = All Tangible and Intangible Asset Created through the Investment
$BVA_{as\,D}$ = Book Value of Asset a in space layer s recorded at date D

The new asset impact of the investment or cash flows considers all expenditure into the creation of new assets through the investment. The formula measures the value of all newly created assets across all space layers. The multiple sigma operators ($\sum$) serve the purpose of identifying and summing the different Assets created in different space layers. Time is relevant here in terms of when or in which period the asset is created and added on the books.

Measuring and accounting for new asset impact is critically important when considering the space impact of outer space projects. This is because outer space projects may be highly risky, with distant returns, but they may be creating new assets through exploration, settlement, and mining. The coefficient k_a is a coefficient of transition, and it is defined on the individual asset level. It is a coefficient with a value between $-1 \leq k_a \geq 1$ and it can be used to qualify the assets as the absolute value of new assets does not always reflect the future value of the asset given economy wide transformations. For example, a new coal mine and a new space-based solar power generation plant cannot be valued the same way. The former will most likely be out of commission in the near future, while the latter will grow in importance and value.

13.1.2.6   New Money Impact

$$NMI_T = (II \times DR \times BLR) + mm \times (II + X_T - M_T) \qquad (13.15)$$

mm = Money Multiplier
DR = Debt Ratio
BLR = Bank Loan Ratio
M_T = Planned Imports
X_T = Expected Exports
II = Initial Investment
N = Total Years in the Investment

The new money impact of the investment or cash flows considers the new money creation impact of the investment. All investments could be new money through debt or may create new money through the banking system (given our current model). As all banknote money is confined to one space layer and given that money through the datasphere is still managed by the same logic, there is no sigma to measure and add the impact across layers. Similarly, given that the formula measures the overall new money impact of the investment, there is no sigma that adds the impact across the years.

Measuring and accounting for new money impact is necessary to address the fact that all investments contribute to our existing monetary architecture through the debt-based system. As such, with large investments, this is an important aspect to consider.

13.1.3   Integrating into Value

The space value framework is not just about quantifying the space impact of cash flows, it is also about integrating them into our models and equations of value and return.

Once we have mapped and quantified the multilayered space impact of investments and cash flows, we can then make sure that negative impacts are prevented and avoided. To do so, negative impacts must be made to affect the value of the cash flows and/or investment. Equation 13.16 is one example of such an integration using the Net Present Value formula. The equation uses the absolute value of the negative space impacts and adds the negative external to the total for theoretical clarity.[1]

$$\text{Negative Impact Adjusted NPV} = -\left|\text{NNSV}_{T,S}\right| - \text{II} + \sum_{t=1}^{T} \frac{CF_t}{(1+r)^t}$$

$$(13.16)$$

$\text{NNSV}_{(T,S)}$ = The Sum of Negative Impacts Across All Years and Space Layers

Similarly, when negative impacts are avoided, the value of the investment can be complemented with the positive Space Value of the cash flows, as in Eq. 13.17, using the example of the firm value equation using Free Cash Flow

---

[1] See Papazian (2022) for additional examples and a more extensive discussion.

to Firm over time.

$$\text{Positive Impact Adjusted Firm Value} = \frac{\text{NSV}_1}{\text{WACC} - J_p} + \sum_{t=1}^{T} \frac{\text{FCFF}_t}{(1+\text{WACC})^t}$$
$$+ \frac{\text{FCFF}_{T+1}}{(\text{WACC} - g).(1+\text{WACC})^T}$$
$$(13.17)$$

Naturally, these are high level examples to demonstrate that through the space value framework, we can do both, eliminate negative impacts and optimise positive impacts across all space layers. I do not discuss the derivation and details of these examples here. They are, along with other non-discounting model examples, discussed at length in Papazian (2023, 2022).

This brings us to the next conceptual contribution of the space value framework, the Space Growth Rate (SPR).

13.1.4   The Space Growth Rate

The Space Growth Rate measures the implied periodic rate of growth using the Initial Investment (II) and the aggregate Net Space Value of the investment across the T periods and S layers of the opportunity or project. In other words, it is the summary rate that when used to compound the initial investment into the future, will give us the total Net Space Value of the investment at maturity.

The Space Growth Rate (SPR) is a summary rate that can be used similarly to how the discount rate is used in our current models. With the space value of money principle, investors can continue to pursue their risk and time adjusted returns, but their impact must be accounted for and compounded into the future when relevant. Figure 13.5 and Eqs. 13.18 and 13.19 reveal the relationship between the space growth rate and the Net Space Value of an investment, where the investment is considered as a series of cash expenditures, and the SPR is used to compound them into the future (Papazian 2022, 100–101).

The Space Growth Rate can be used to set the minimum required space growth rate for public and private investments. Just like benchmark discount rates are used to measure the risk and time value of cash flows and investment opportunities, we could set positive benchmark space growth rates greater than zero that would set the minimum positive space impact required for public and private investments (Papazian 2023, 2022). This is an important,

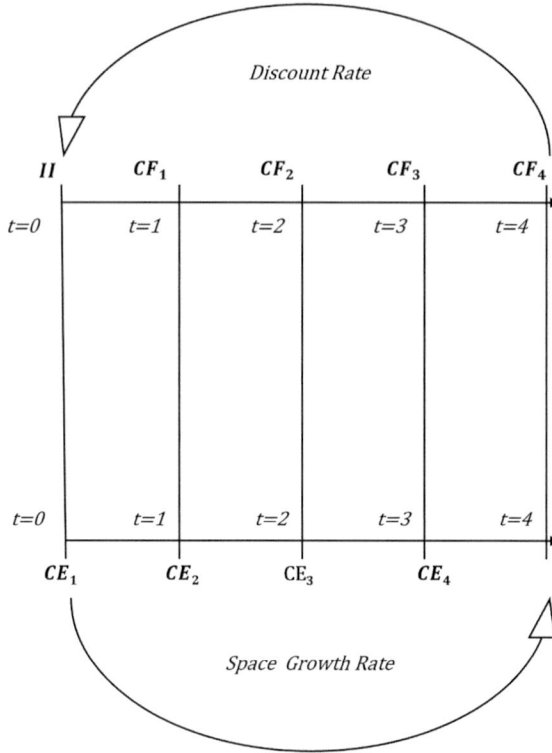

Fig. 13.5 The double timeline and the space growth rate (*Source* Papazian 2022)

and yet missing, tool in our current financial framework and mathematics. Our focus on risk and time excludes the space impact of cash flows as a relevant component of value, and thus, as discussed in earlier sections, discriminates against our evolutionary investments which tend to have very high risks, distant cash flows, and abundant space impact.

$$\mathbf{SPR} = \sqrt[T]{\frac{\mathbf{NSV}_{T,S}}{\mathbf{II}}} - 1 \qquad (13.18)$$

$\mathrm{NSV}_{(T,S)}$ = Net Space Value
SPR = The Space Growth Rate per period
II = Initial Investment
T = All Time Periods involved in the Investment
S = All Space Layers Involved in the Investment

$$II = \sum_{t=0}^{T} CE_t = \text{Initial Investment}$$

$$\mathbf{NSV}_{T,S} = \sum_{t=0}^{T} \mathbf{CE}_t (1 + \mathbf{SPR})^{T-t} \qquad (13.19)$$

13.2 Conclusion

The introduction of the analytical dimension of space and the space value of money principle allows us to correct the current imbalance in finance theory and practice. Risk and time may be of concern to the mortal risk-averse return-maximising investor, but from the perspective of the human collective, our sustainable evolutionary continuity in space takes priority.

The introduction of space and the space value of money principle give us the framework through which we can transform our equations of value and return. Accounting for space impact and integrating it into our models is relevant to our sustainability as well as expansion. Investors can continue discounting their future expected cash flows, as long as they are also accounting and compounding their space impact into the future. Cash flows cannot be assessed independently of the space impact it would take to achieve or expect them.

The space-adjusted equations I propose allow us to integrate space impact into our models and redesign our valuations to account for our footprint. Indeed, our financial value models must be able to properly price our evolutionary investments. We need models that can, at least in principle, result in a positive valuation for our evolutionary investments. Massive investments in the present that carry incredibly high risks, very distant returns, and high positive impact on space must become investable, and only a transformed framework and mathematics can make this possible.

Once we have transformed and reformed our financial value framework and mathematics, we can liberate our investment flows from the claws of risk and time. This will allow the proper valuation of outer space development programmes and investments. While this is a necessary transformation, it is not sufficient.

In the next chapter, I discuss the implications of a space-adjusted framework and mathematics for our monetary architecture.

References

Papazian, A. 2023. *Hardwiring Sustainability into Financial Mathematics: Implications for Money Mechanics*. New York: Palgrave Macmillan. https://doi.org/10.1007/978-3-031-45689-3.

Papazian, A. 2022. *The Space Value of Money: Rethinking Finance Beyond Risk and Time*. New York: Palgrave Macmillan. https://doi.org/10.1057/978-1-137-59489-1.

14

Monetising Space

None of the paradigms that define us here on earth - the borders, the parochialism, the divide, should mar our presence in space.
Rakesh Sharma, Soyuz T-11 Astronaut, 1984

I chased my dreams.
Guion Bluford, STS-8 Challenger Astronaut, 1983

The previous chapters introduced space into finance, as analytical dimension and our physical context, established a new and complementary principle of value, the space value of money, and offered a space-adjusted financial mathematics. These transformations achieve respect for space and provide the tools through which we can value and invest in space, and thus outer space. Space impact, heretofore abstracted and left out, is now an integral part of our equations of value and return.

These changes are needed and necessary to allow the proper valuation of our evolutionary investments, and to remove the bias against highly risky and distant return investments. The preferences of the mortal risk-averse return-maximising investor can now be balanced with the priorities of the species and the planet. Indeed, beyond Earthly money supply, our impact on and in space can now be valued and considered.

The next logical step is to apply the transformed framework and mathematics to our monetary architecture. In other words, if investors must abide by the space value of money principle and consider their space impact, then so must money creators. As we discussed in Chapter 7, our current debt-based

© The Author(s), under exclusive license to Springer Nature
Switzerland AG 2024
A. V. Papazian, *Financing the Race to Space*,
https://doi.org/10.1007/978-3-031-73102-0_14

monetary architecture is spaceless. This is why a space-adjusted financial value framework and mathematics are necessary prerequisites to this discussion.

The muzzle, leash, and whip in space, born out of the main features of debt and calendar time, can be transcended when we use *space value creation* as the foundational logic of money creation. In the following discussion I offer a solution by way of a new money creation channel based on space value creation (Papazian 2022, 2023). Ultimately, the solution is about an instrument improving a system.

14.1 Money Creation Through Space Value Creation

The alternative logic of money creation I propose is built upon the introduction of a new instrument for the purpose. In other words, the proposed transformation can be implemented with minimal systemic shocks and with relative ease and speed.

To remove any doubts from the reader's mind, it is entirely possible to improve and fine-tune our money creation methodology. Furthermore, it is necessary to address the many evolutionary challenges we have created for ourselves. Calendar time, monetary gravity, and monetary hunger undermine a sustainable and expansive human productivity. They are prohibitive forces on our journey in outer space. Thus, we should and must consider this avenue as an opportunity for innovation and improvement in money creation.

> [I]f the Bank of England can create and back banknotes by a deposit in the banking department of the Bank of England, if the Bank of England can create new money by loaning to its own wholly owned subsidiary, if the Federal Reserve can create new money by buying toxic Collateralised Debt Obligations and Mortgage-Backed Securities or by buying commercial paper, there is no reason why they cannot back or create new money through an alternative equity-like instrument that shares risks, shares the ownership of the assets created through the instrument, has a tangible and inspiring positive space impact, and helps resolve our evolutionary challenges. (Papazian 2022, 223)

The balance sheet of the issue department of the Bank of England (Table 14.1) reveals that the British Pounds in circulation are backed by government debt securities and an internal deposit at the banking department of the Bank of England.

During the 2007/8 financial crisis and the 2020 pandemic the Bank of England injected hundreds of billions of new money or liquidity into the

Table 14.1 BOE balance sheet, issue department, in (£mn), As of Feb 2023

	2023	2022
	£mn	£mn
Assets		
Securities of, or guaranteed by, the British Government	1,536	1,698
*Other securities and assets including those acquired under reverse repurchase agreements	84,371	84,742
Total Assets	**85,907**	**86,440**
Liabilities		
Note Issued		
In Circulation	85,907	86,440
Total Liabilities	**85,907**	**86,440**
*		
Other securities and assets including those acquired under reverse repurchase agreements		
Deposit with Banking Department	84,261	82,387
Reverse repurchase agreements	110	2,355
	84,371	**84,742**

Source Bank of England (2023)

financial system by buying debt securities through loans to its own wholly owned subsidiaries.[1]

> When the APF is used for monetary policy purposes, purchases of assets are financed by the creation of central bank reserves…. The APF transactions are undertaken by a subsidiary company of the Bank of England – the Bank of England Asset Purchase Facility Fund Limited (BEAPFF). The transactions are funded by a loan from the Bank…. (Bank of England 2021, 117)

In Chapter 7 I discussed the nature and content of the bank's QE strategy designed to inject new money into the economy through debt instruments and transactions. The Bank of England created the *Bank of England Asset Purchase Facility Fund Limited* (BEAPFF) on the 30th of January 2009, a wholly owned subsidiary, a private limited company, with Companies House registration number 06806063 (Companies House 2024), which is still active and has been used since then to execute the bank's monetary policy, i.e., Quantitative Easing (QE) strategy. This is clearly stated in the description of business activity in the company's annual report and accounts (BEAPFF 2023). In its March 2022–February 2023 annual report and accounts BEAPFF states that "the balance sheet of the company totalled

---

[1] See Gillian Tett (2010) for a fascinating account of the 2007/8 financial crisis. See Lewis (2010) and Lowenstein (2010) for parallel accounts of the same.

£846.2 billion as at 28 February 2023 (2022: £895.3 billion). The Company's principal liability was the loan from the Bank [of England] of £843.7 billion (2022: £894.9 billion) (BEAPFF 2023, 13).

In other words, if we can create money through an internal deposit and/or through loans to wholly owned subsidiaries, or by purchasing debt instruments, then we can also invent money with other alternative equally credible instruments and entities. What I propose here and have done so in Papazian (2022, 2023) is a systemic change that amounts to the introduction of a new financial instrument and mechanism within our existing monetary architecture.

The proposition is to use a set of equally credible institutions and reliable legal frameworks but change the logic of creation. This can be achieved through the introduction of a new instrument, Public Capitalisation Notes (PCNs), which have a different logic and different locus of injection. They differ from debt instruments used by the Federal Reserve, Bank of England and the European Central Bank (Bernanke 2009). Echoing the commonly used policies, Quantitative and/or Credit Easing (QE/CE), I call this strategy Value Easing (VE).

14.2   The Instrument: Public Capitalisation Notes (PCNs)

The logic and purpose of Public Capitalisation Notes are designed to help us transcend the three systemic bottlenecks created by debt-based money. This is possible to achieve through PCNs because they are designed and considered within a transformed space-adjusted framework. They can deliver practical change in our monetary architecture thanks to the parallel and already introduced changes in our financial value framework and mathematics. Unlike debts, which are valued in risktime, PCNs are valued in space.

PCNs are conceived as instruments that can be used by any central bank. They are not debt instruments and introduce a new logic of money creation founded on space value creation. In other words, if with debt instruments the trigger of money creation is the agreement to repay, with PCNs, this is changed to a commitment to create necessary positive space value and share the returns when and if they occur.

Looking at the US Federal Reserve balance sheet (Table 14.2), we can see the prominent and widespread use of credit facilities, loans, and debt instruments on the asset side. The suggestion here is to add PCNs on this balance sheet, again on the asset side. I propose a NASA PCN as an architectural innovation that can empower the US outer space sector.

In 2015, the US Congress passed resolution H.R.1508, the Space Resource Exploration and Utilization Act, which declared clearly the intention to:

Table 14.2 Assets, liabilities, and capital of the Federal Reserve System ($ billions)

	March 27, 2024	September 27, 2023
Total assets	**7,485**	**8,002**
Securities held outright	7,009	7,440
U.S. Treasury securities	4,618	4,958
Federal agency debt securities	2	2
Agency mortgage-backed securities	2,388	2,480
Repurchase agreements	0	0
Foreign official	0	0
Other	0	0
Loans	142	198
Discount window	6	3
Bank Term Funding Program	133	108
Paycheck Protection Program Liquidity Facility	3	5
Other credit extensions	0	82
Net portfolio holdings of Main Street Facilities LLC	15	19
Net portfolio holdings of Municipal Liquidity Facility LLC	0	6
Net portfolio holdings of Term Asset-Backed Securities Loan Facility II LLC	0	1
Central bank liquidity swaps	0	0
Other assets	319	337
Total liabilities	**7,442**	**7,959**
Federal Reserve notes	2,293	2,273
Deposits held by depository institutions other than term deposits	3,472	3,169
Reverse repurchase agreements	873	1,755
Foreign official and international accounts	354	312
Others	518	1,443
U.S. Treasury, General Account	772	672
Treasury contributions to credit facilities	7	13
Other liabilities	24	77
Total capital	**43**	**43**

Source Fed (2024)

[f]acilitate the commercial exploration and utilization of space resources to meet national needs; discourage government barriers to the development of economically viable, safe, and stable industries for the exploration and utilization of space resources in manners consistent with the existing international obligations of the United States; and promote the right of U.S. commercial entities to explore outer space and utilize space resources, in accordance with such obligations, free from harmful interference, and to transfer or sell such resources. (US Congress, 2015)

To go beyond legislative support, Fig. 14.1 describes a possible NASA PCN that could be used by the US Treasury and the US Federal Reserve to jump start a massive investment drive in public and private outer space development and exploration through NASA. Naturally, NASA is given as an example, and other alternative viable institutions can also be used.

PCNs differ from debts in a variety of ways. First of all, they are not debt instruments (addresses monetary hunger). Moreover, they have no-maturity (addresses monetary gravity and calendar time) and they are equity-like (shares risks and assets and returns when due) and have very high space value/impact (addresses challenges faced by outer space exploration and settlement projects).

When issued by qualified government agencies in collaboration with the private sector (Public Private Partnerships) and under Treasury sponsorship, they become viable instruments that the Federal Reserve can purchase. They increase the central bank's balance sheet (like Quantitative Easing) but inject new liquidity outside the banking system (unlike Quantitative Easing) (Papazian, 2022, 2023).

The created money will of course be in bank accounts within the banking system. The difference here is that the new money created through a NASA PCN is allocated to actual and immediate spending on innovation, technological advancement, and new asset creation. Unlike bonds purchased for QE, the impact on output and productivity is immediate. This ensures that PCNs are far less inflationary. Moreover, PCNs do not condition the space impact of the created new money by new or further lending by banks, given their increased liquidity under QE.

In the case of QE, the initial point of injection being the purchase of debt instruments, the eventual impact of the new liquidity is dependent on further bank lending, which, as discussed in previous chapters, is guided by and built upon a risk and time-focused value framework. PCNs, because of their design and structural features, allow the funding of our many evolutionary challenges and necessary economy wide investment programs without any additional debt. Moreover, they transcend the risktime framework as they

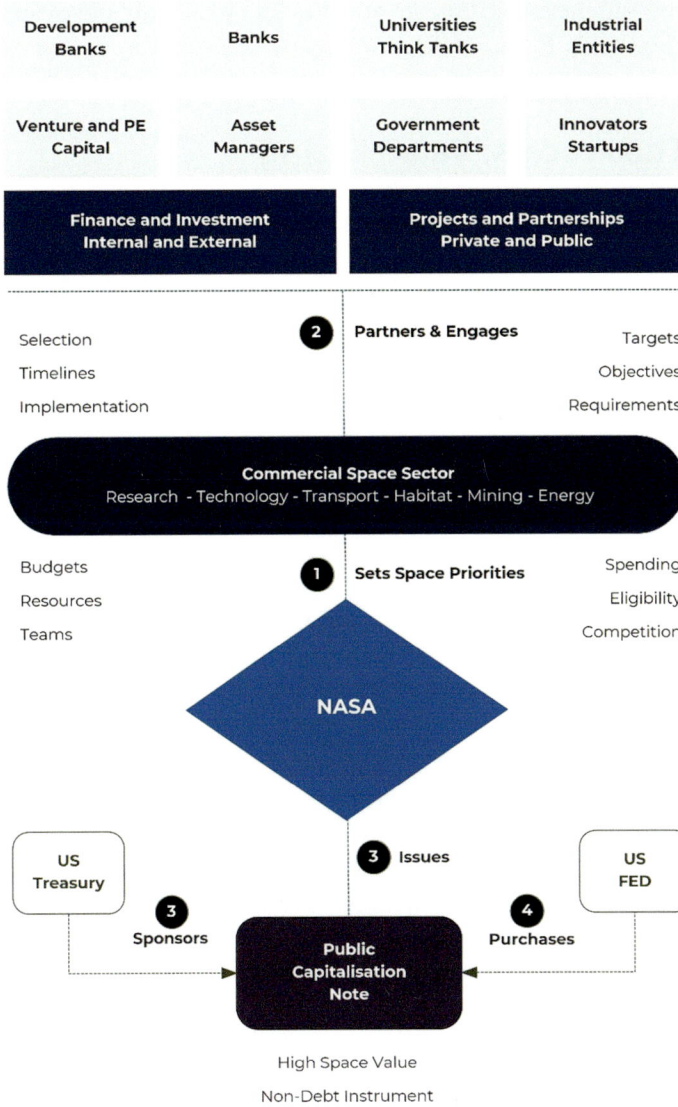

Fig. 14.1 NASA PCN (*Source* Adapted and updated from Papazian [2022, 2023])

create space value immediately and can by design address opportunities with high risks and distant returns.

The transformations proposed here introduce a new logic of money creation and a new locus of injection that allow us to address the challenges of debt-based money and unleash the massive investment programs aimed at achieving our next big leap in outer space. I call this novel approach of monetisation Value Easing (VE).

14.3   The Process: Value Easing

Value Easing (VE) is the transactional process undertaken by a central bank that consists in purchasing non-debt, no-maturity, equity-like, high space impact, value creating instruments from qualified government agencies and/ or public private partnerships (PPP), with relevant Treasury sponsorship, that increases the central bank's balance sheet and injects new liquidity outside the banking system (Papazian 2023). Naturally, the created new digits, or money, are in the form of central bank reserves. The point here is that they are injected in the account of a NASA led Public Private Partnership that will spend it to boost the US outer space economy, and not in the accounts of previous bondholders who may lend it out to new borrowers.

Value Easing is less inflationary than Quantitative Easing and Credit Easing, and it allows the injection of new liquidity to be directed where it is most necessary and needed, instead of being left to the lending priorities and preferences of bondholders and banks. This is necessary given the risk and time-based analytical framework within which banks create and manage their assets and liabilities. Moreover, directing new liquidity to clearly identi-fied output creating and productivity enhancing investments ensures that the new injection creates value, and does not simply lead to asset price inflation.

Value Easing can be used to address our evolutionary challenges, like outer space exploration and settlement, because it is built on the very premise of positive space value impact. Moreover, PCNs and Value Easing, unlike debts and quantitative easing, can deal with high risk levels and distant return time-lines. Moreover, Value Easing allows us to monetise our footprint in space. Specifically, in outer space, it allows us to monetise the value we will create on the moon, Mars, and anywhere else we may end up going on our journey.

In practical terms, the introduction of Public Capitalisation Notes and Value Easing transforms the funding sources of government expenditures. Alongside debts and taxes, PCNs introduce a new channel that can be directed towards the outer space sector. Fig. 14.2 reflects this diversification on the government sector side. It also shows the impact on private sector revenues as PCNs are conceived for country wide private and public sector partnerships.

Through subcontracting and Public Private Partnerships, PCNs also feed into private sector revenues and allow a wider distribution and utilisation

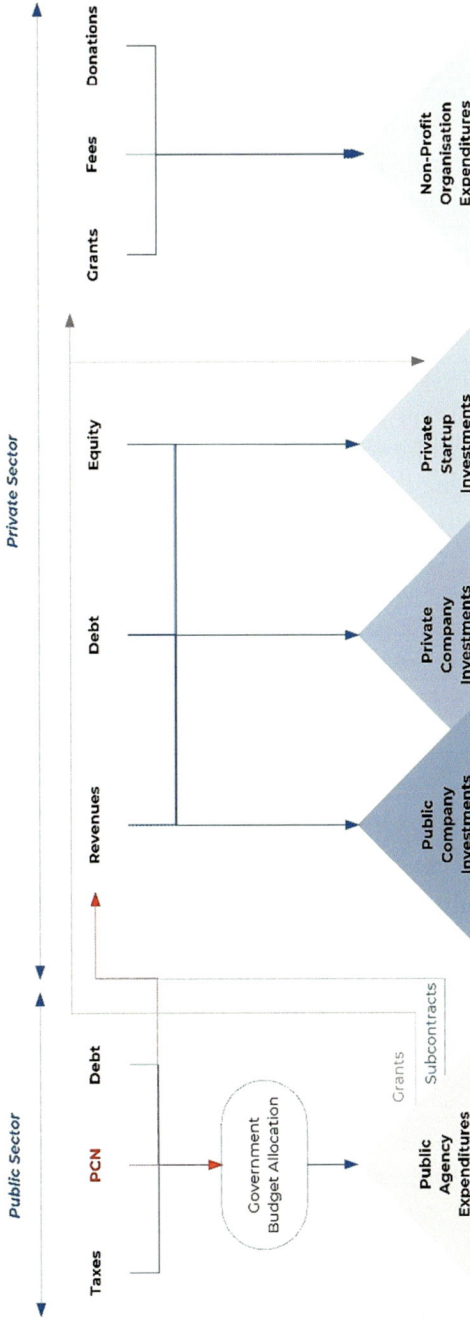

Fig. 14.2 PCNs and funding sources and investment flows in outer space sector (*Source* Author)

of the funds for a unified mission: breakthroughs in outer space aimed at inventing, manufacturing, deploying, and maintaining the new habitats on the moon and Mars.

A NASA PCN can allow the US public and private outer space sectors to go beyond the limitations imposed by our current framework and architecture. While doing so, they can introduce a much-needed change in public expenditure financing structures. Indeed, given US National Debt levels, discussed in detail in Chapter 15, this is a necessary transformation.

14.4   Conclusion

The systemic challenges posed by a debt-based monetary architecture, imposing limits on human productivity, and acting as a muzzle, leash, and whip in space, can be addressed through the above-proposed instrument and process. PCNs and Value Easing grant us the opportunity to transcend those systemic bottlenecks. In a document explaining the logic of QE, Bank of England states:

> The money we used to buy bonds when we were doing QE did not come from government taxation or borrowing. Instead, like other central banks, we can create money digitally in the form of 'central bank reserves'.... We use these reserves to buy bonds. Bonds are essentially IOUs issued by the government and businesses as a means of borrowing money. (Bank of England 2024)

Now, imagine one day soon reading the below in a Bank of England or US Federal Reserve policy paper. The theoretical and practical bridge that makes the below possible is the space-adjusted financial value framework and mathematics discussed in previous chapters.

> The money we used to buy PCNs when we were doing VE did not come from government taxation or borrowing. Instead, like other central banks, we can create money digitally in the form of 'central bank reserves'.... We use these reserves to buy PCNs. PCNs are essentially financial instruments issued by the government or agencies as a means of investing in an economy wide investment program that brings public and private sector entities together and aims at addressing our most daunting national and evolutionary challenges.

We must monetise space to be able to transcend the limitations of debt-based money, a conceptual tool built entirely on the monetisation of risk and time.

References

Bank of England. 2024. Quantitative Easing. The Bank of England. https://www.bankofengland.co.uk/monetary-policy/quantitative-easing. Accessed 25 June 2024.

Bank of England. 2023b. Bank of England Annual Report and Accounts 22/23. Bank of England. https://www.bankofengland.co.uk/-/media/boe/files/annual-report/2023/boe-2023.pdf#page=201. Accessed 1 February 2024.

Bank of England. 2021. Bank of England Annual Report and Accounts 20/21. Bank of England. https://www.bankofengland.co.uk/-/media/boe/files/annual-report/2021/boe-2021.pdf#page=97. Accessed 2 February 2024.

BEAPF. 2023. Annual Report and Accounts for Bank of England Asset Purchase Facility Fund Limited. Companies House. https://find-and-update.company-information.service.gov.uk/company/06806063/filing-history. Accessed 12 May 2024.

Companies House. 2024. Bank of England Asset Purchase Facility Fund Limited. Companies House. https://find-and-update.company-information.service.gov.uk/company/06806063. Accessed 12 May 2024.

FED. 2024. Federal Reserve Balance Sheet Developments. US Federal Reserve. https://www.federalreserve.gov/publications/files/balance_sheet_developments_report_202405.pdf. Accessed 12 June 2024.

Lewis, M. 2010. *The Big Short*. W. W. Norton & Company.

Lowenstein, R. 2010. *The End of Wall Street*. Penguin Books.

Papazian, A. 2023. *Hardwiring Sustainability into Financial Mathematics: Implications for Money Mechanics*. New York: Palgrave Macmillan. https://doi.org/10.1007/978-3-031-45689-3.

Papazian, A. 2022. *The Space Value of Money: Rethinking Finance Beyond Risk and Time*. New York: Palgrave Macmillan. https://doi.org/10.1057/978-1-137-59489-1.

Tett, G. 2010. *Fool's Gold: How Unrestrained Greed Corrupted a Dream, Shattered Global Markets and Unleashed a Catastrophe*. Abacus.

US Congress. 2015. H.R.1508, the Space Resource Exploration and Utilization Act. United States Congress. https://www.congress.gov/bill/114th-congress/house-bill/1508. Accessed 12 January 2024.

Part V

Institutional and Policy Transformations in Space

This part explores two actual example case studies discussing the side benefits of adopting the space value framework and concludes the book. The first example is a US wealth floor replacing a debt ceiling, and the second example is a UK New Outer Space Deal replacing years of austerity and underinvestment.

15

US Debt Ceiling to Wealth Floor

Space is going to be commonplace.
Christa McAuliffe, Teacher and STS-51-L Challenger Astronaut, 1986

Although I dreamed of flying from childhood, that was utterly unthinkable. I had to start work early, a shoeshine boy — a poor man's profession — or selling vegetables.
Arnaldo Tamayo Méndez, Soyuz 38 Cosmonaut, 1980

A space-adjusted financial value framework, financial mathematics, and monetary architecture are key for unlocking the resources we need to fund our expansion in outer space. They are necessary in order to transcend the existing biases against long horizon and highly risky projects and investments. The integration of space impact and responsibility into our equations allows an entirely different approach to valuation that goes beyond the risk and time value of cash flows and considers their space impact to be equally relevant. This, in turn, allows us to transcend the preferences of the mortal risk-averse return-maximising investor and consider the priorities of the species and the planet.

All of the above facilitate the introduction of new money creating instruments, Public Capitalisation Notes (PCN), which help us overcome the limitations of a purely debt-based monetary architecture. PCNs remove the constraints imposed by calendar time, monetary gravity, and monetary hunger. They allow us to overcome the restrictions imposed by our own conceptually imposed muzzle, leash, and whip in space.

© The Author(s), under exclusive license to Springer Nature Switzerland AG 2024
A. V. Papazian, *Financing the Race to Space*,
https://doi.org/10.1007/978-3-031-73102-0_15

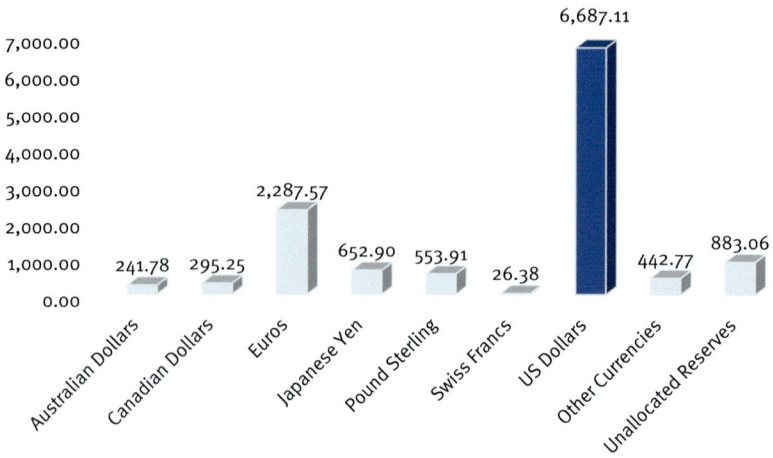

Chart 15.1 Global currency reserves percentages 2023, in USD billions equivalent (*Source* IMF [2024])

Once we have introduced the analytical dimension of space into finance, established the principle of space value of money, adjusted our equations to integrate space impact, and created the commensurate instruments of money creation, we can begin the funding of our evolutionary investments.

In this chapter I discuss an important side-benefit of this transformation within the context of the US national debt and debt ceiling or limit. While this is specifically a US challenge, given the role of the US Dollar in the world economy, it is a global bottleneck for the world financial system. Despite the ambitions of many to compete and/or replace the US dollar, the dollar remains the most commonly used global reserve currency today (IMF 2024) (Chart 15.1). Moreover, the discussion applies to other countries as well, even if they do not have a statutory debt limit on public borrowing.

A space-adjusted financial and monetary framework allows us to free our monetary potential from debt and the many limitations it imposes. Even in its most limited form, Value Easing, or money creation based on space value creation, can help us balance the existing entirely debt-based system. This is because it can be used to change the logic of money creation for the amounts that are already being created. In other words, even if we do not fund massive investment programs for outer space exploration and settlement, we can still transform the logic of the money we are already creating at the moment— replacing the debt logic with a space value creation logic.

The main purpose of this chapter is to discuss how Public Capitalisation Notes can be used to transform the US debt ceiling into a wealth floor.

15.1 From US Debt Ceiling to Wealth Floor

15.1.1 US National, Federal, Public Debt

The US treasury uses national, federal, and public debt interchangeably. Thus, I follow the same approach here. To start off, the US Treasury's definition of US public debt is the right place to start.

> The national debt is the amount of money the federal government has borrowed to cover the outstanding balance of expenses incurred over time. In a given fiscal year (FY), when spending (ex. money for roadways) exceeds revenue (ex. money from federal income tax), a budget deficit results. To pay for this deficit, the federal government borrows money by selling marketable securities such as Treasury bonds, bills, notes, floating rate notes, and Treasury inflation-protected securities (TIPS). The national debt is the accumulation of this borrowing along with associated interest owed to the investors who purchased these securities. As the federal government experiences reoccurring deficits, which is common, the national debt grows. (US Treasury 2024a)

In Chart 15.2, we can see the historical debt outstanding of the US at the end of each fiscal year from 1791 to 2023.

Public debt in the US has been subject to a statutory limit since 1917. The US Congress sets this limit. Only 0.5% of total debt is excluded from the debt

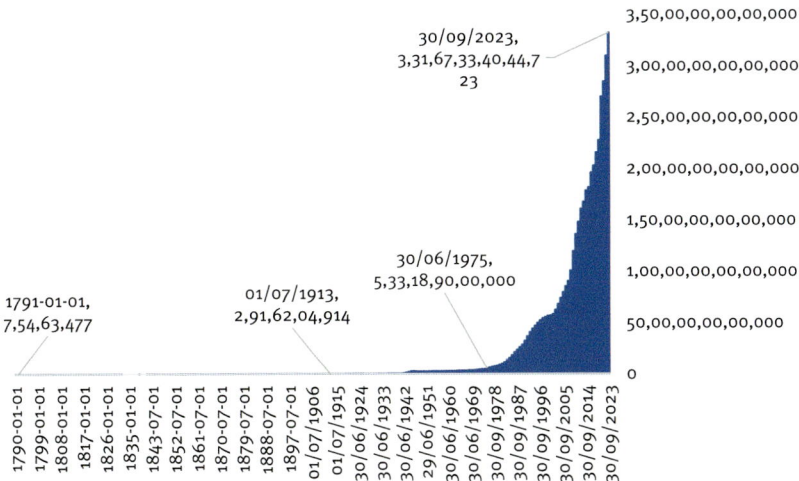

Chart 15.2 US government historical debt outstanding 1790–2023 (*Source* US Treasury [2024b])

limit (Austin 2015). The US Treasury defines total public debt outstanding and debt subject to limit as follows:

> The Total Public Debt Outstanding represents the total par (principal amount) of Treasury marketable securities and Treasury non-marketable securities currently outstanding. Treasury Marketable securities include Treasury bills, notes and Treasury Inflation-Protected Securities (TIPS), all of which can be bought and sold in the secondary market at prevailing market prices. Treasury Non-marketable securities include savings bonds as well as special securities issued only to state and local governments and Federal trust funds such as Social Security…
>
> The Total Public Debt Subject to Limit is the maximum amount of money the Government is allowed to borrow under authority granted by Congress.
>
> Total Public Debt Subject to Limit is defined as the Total Public Debt Outstanding less the Unamortized Discount on Treasury Bills and Zero-Coupon Treasury Bonds, old debt issued before 1917, and old currency called United States Notes, as well as Debt held by the Federal Financing Bank and Guaranteed Debt (US Treasury, 2024c).

The debt limit or ceiling has been a central institutional bottleneck and the subject of intense debate in the US. A detailed and authentic account of the nature and history of the debt limit can be found in Austin (2015).

> Congress has always restricted federal debt. The Second Liberty Bond Act of 1917 included an aggregate limit on federal debt as well as limits on specific debt issues. Through the 1920s and 1930s, Congress altered the form of those restrictions to give the U.S. Treasury more flexibility in debt management and to allow modernization of federal financing. In 1939, a general limit was placed on federal debt…. Federal debt accumulates when the government sells debt to the public to finance budget deficits and to meet federal obligations or when it issues debt to government accounts, such as the Social Security, Medicare, and Transportation trust funds. (Austin, 2015, 1)

In the first half of 2023, the US Treasury reached its debt limit of $31.4 trillion (Edelberg and Sheiner 2023) and the ceiling was raised on the 31st of May 2023. On the 3rd of May 2023, in a blog that reads as a warning, the White House wrote:

> New analyses by both the Congressional Budget Office and the U.S. Department of the Treasury suggest the United States is rapidly approaching the date at which the government can no longer pay its bills, also known as the "X-date." History is clear that even getting close to a breach of the U.S. debt ceiling could cause significant disruptions to financial markets that would

damage the economic conditions faced by households and businesses. Real time data, shown below, indicate that markets are already pricing in political brinkmanship related to Federal government default through higher risk premia. (White House, 2023)

The x-date, which is when the Treasury runs out of funds, is a hypothetical date that will hypothetically be avoided ad infinitum. However, irrespective whether this is possible or not, it further illustrates the limitations of the debt-based monetary architecture. All the systemic challenges discussed in previous chapters are further emphasised by the very existence of the x-date. Extending the leash does not remove the limitations of the leash. It simply highlights the existence of the leash and the need to extend it.

A space-adjusted financial value framework, mathematics, and monetary tools can help us address the limitations imposed through the very existence of the debt limit or ceiling. They can help us reinvent the debt ceiling into a wealth floor.

15.1.2 From Debt Ceiling to Wealth Floor

Adopting the main propositions of this book can transform the US Debt Ceiling into a Wealth Floor. As introduced in the previous chapter, the monetisation of space through Value Easing using PCNs provides us the tools and opportunity to transform the portfolio of instruments we use for money creation, and to further diversify the sources of funding we use to support public expenditures. This is how the NASA PCN can fund massive investment programs dedicated to inventing, manufacturing, deploying, and maintaining the new habitats of the future.

In parallel, the same logic can be used to also expand the portfolio of instruments used to fund other public expenditures. Instead of issuing debt instruments, the US Treasury can use Public Capitalisation Notes (PCNs) to fund a variety of large investment programs which can reduce and replace the issuance of new debt over time. This is how the debt ceiling can be transformed into a wealth floor, through the gradual replacement of debt instruments with PCNs. This does not mean that government T-bills or bonds should be removed from market, they can still be used to provide benchmark rates to the market. The difference here is that they will not be the only kind of instrument issued by the Federal Government.

Indeed, PCNs can also be used to fund other public expenditure programs. A closely linked field and one example is public education. A sector that is at the foundation of our multi-habitat future. In the next section I give the rationale of a specially designed PCN that could be used as part of an overhaul of the US education system, aimed at bringing equality across all states, while shifting from debt to PCN funding. The space value impact of such a PCN is primarily on human capital development.

15.1.3   USEE PCN: US Educational Equality PCN

Educational equality has a significant history in the United States. Since the Civil Rights Act of 1964, the Equal Educational Opportunities Act of 1974, the Equality Act of 2021, the topic has been continuously debated within American society. All branches of government, legislative, executive, and judicial, have had to address the challenges of educational equality. Indeed, the topic of educational inequalities became even more severe during the 2020 pandemic, revealing resource and capability gaps across schools and districts.

The *USEE PCN* I propose here addresses inequality from a very different perspective. Before going into the details, let me remind the reader the main objective of the existing legislation.

The Equal Educational Opportunities Act of 1974 addresses the issue as follows.

> Equal Educational Opportunities Act - Declares it to be the policy of the United States that all children enrolled in public schools are entitled to equal educational opportunity without regard to race, color, sex, or national origin; and that the neighborhood is the appropriate basis for determining public school assignments. States that the purpose of this Act is to authorize concentration of resources under the Emergency School Aid Act on educationally deprived students and to specify appropriate remedies for the orderly removal of the vestiges of the dual school system. (US Congress, 1974)

The educational inequality I propose to address through a PCN is focused on the level of investment across states. This is a high space value opportunity focused on the human capital impact of such an investment. According to an April 2024 press release by the US Census Bureau, based on data from the 2022 Annual Survey of School System Finances, the US witnessed the "largest year-to-year increase in over 20 years for public school spending per pupil."

> Total expenditure by public elementary-secondary school systems totaled $857.3 billion in FY 2022, up 7.8% from the prior year. Of the total expenditure for elementary and secondary education, current spending made up

$746.9 billion (87.1%) and capital outlay made up $84.2 billion (9.8%). (US Census, 2024a)

Looking closely at the total expenditure figures per state, we observe a very unequal distribution, both in terms of overall figures as well as per pupil figures. Figure 15.1 and 15.2 depict these differences. Note that the figures do not take into account inflation and/or geographic differences in cost of living. The highest spending per pupil is in New York, at $29,873, and the lowest level of spending per pupil is in Utah, at $9,552.

The difference between the top and bottom states is $20,321. This is an enormous difference, and it is the kind of inequality that the federal government can address through a PCN that funds major capital expenditure in education across states. Looking at Chart 15.3, we observe the percentage of revenues received through federal, state, and local sources for each state—on average the revenues split is 14.63% from federal sources, 47.62% from state sources, and 38.78% from local sources. This is the revenue that makes the educational expenditure possible. We observe, once again, a diverse set of circumstances.

Vermont, for example, spends $2,256,516,000 (amount is in thousands in Fig. 15.1) and gets 87.4% of its educational revenue/expenditure through state sources, 10.5% from federal sources, and 2.1% from local sources. Meanwhile, Montana, again as an example, spends $2,336,380,000 and gets 20.9% from federal sources, 39.6% from state sources, and 39.5% from local sources. You can find a detailed breakdown of each source in US Census (2024b).

The diverse picture in terms of revenue sources across states is further emphasised when we look at the relative sizes of the federal and state contributions. Figure 15.3 shows the proportion of Federal Sources divided by State Sources across all states. The purpose is to measure and understand their relative size. Once again, the picture is diverse and unequal.

In other words, total expenditure in public elementary and secondary schools, per pupil expenditure of current spending, and federal contribution to state expenditure reveal an uneven and unequal reality. This, naturally, affects all students whatever their race, colour, faith, sexual orientation, and ethnic origin.

A US Educational Equality PCN can aim to ensure that all states are spending at least as much as the current average (Average = $15,841). To achieve this, to raise the per pupil expenditure across all the states to the level of the current average, the Federal Government can issue a Public Capitalisation Note that aims to upgrade and enhance the educational system across the board. Naturally, I have chosen the current average as a hypothetical target

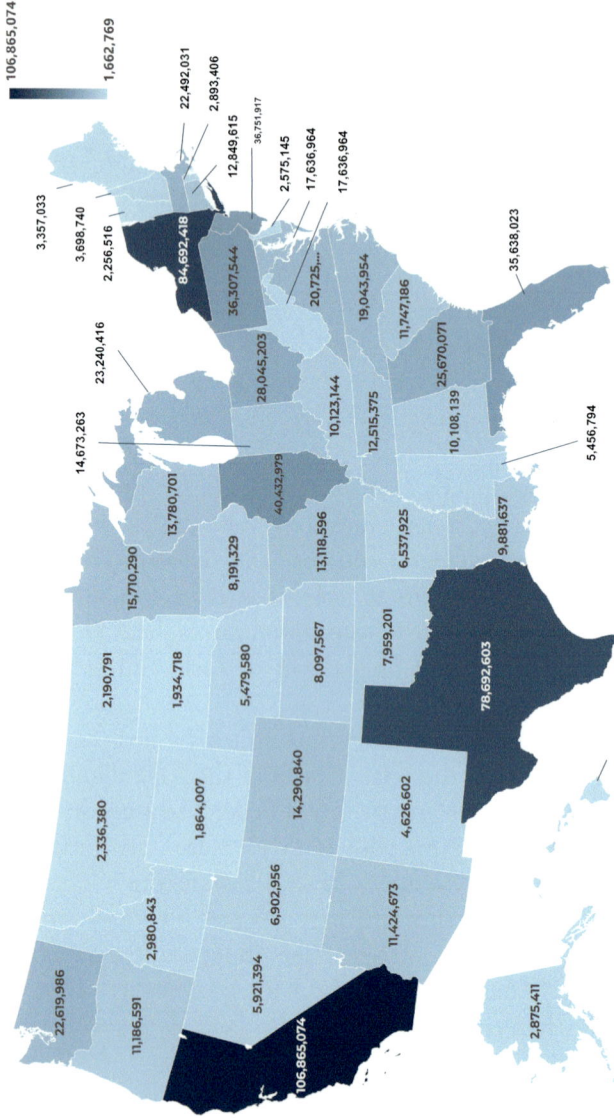

Fig. 15.1 Total expenditure in public elementary and secondary schools, in thousands of US dollars: 2022 (*Source* Author based on US Census [2024b])

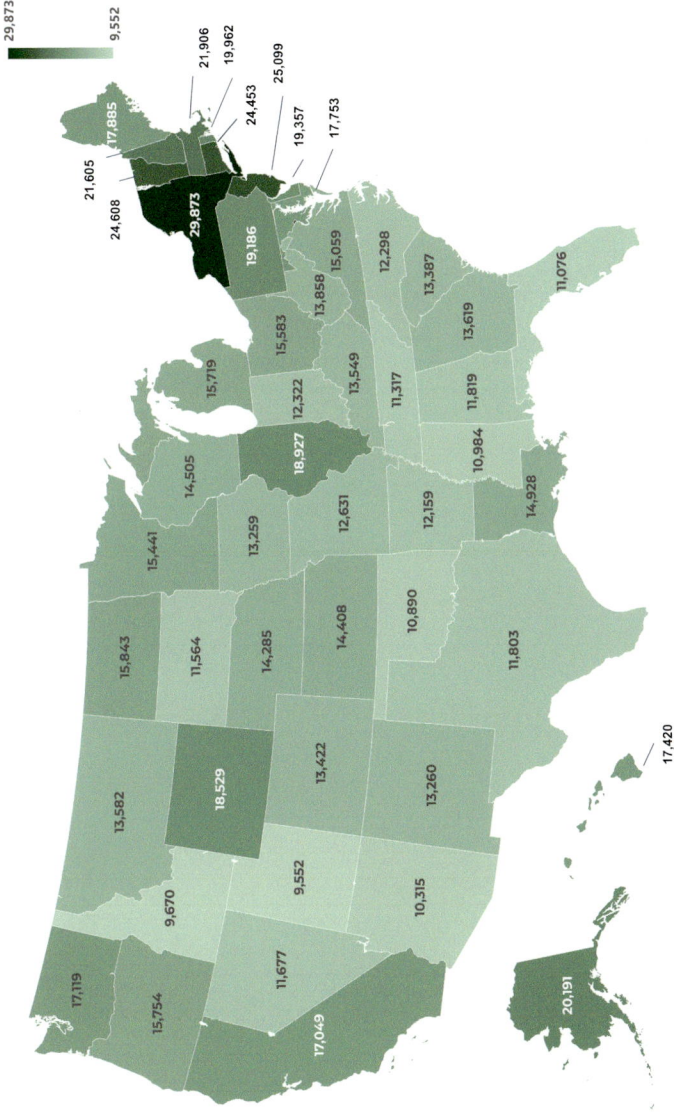

Fig. 15.2 Per pupil expenditure of current spending, US dollars: 2022 (Source Author based on US Census [2024b])

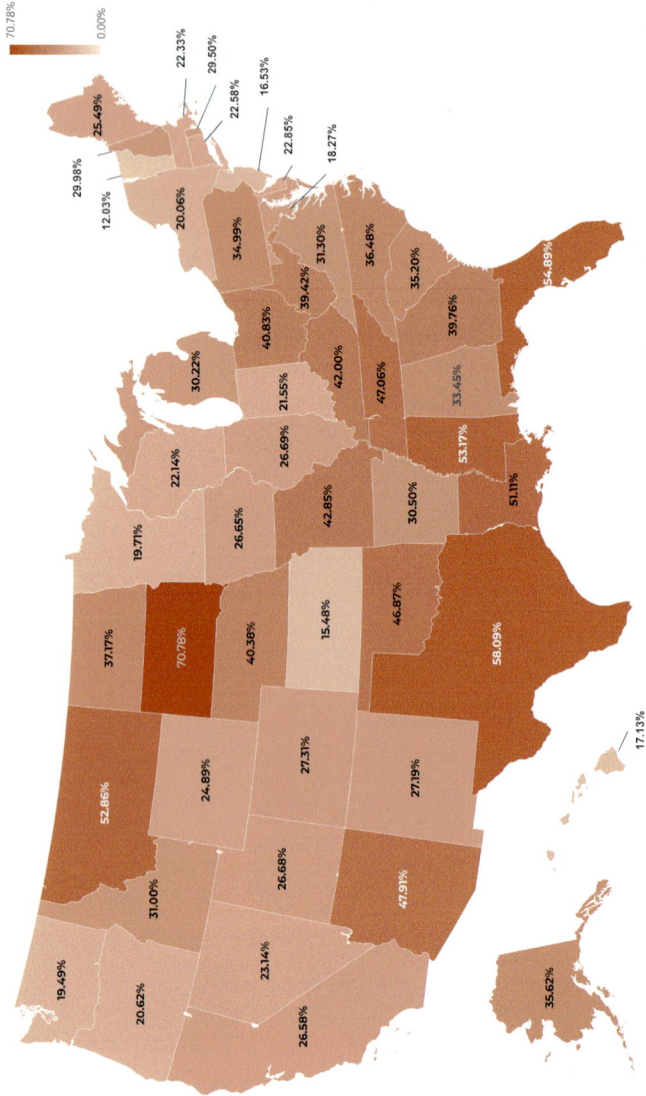

Fig. 15.3 Federal source/state source revenue percentage for public elementary-secondary school systems by state: 2022 (*Source* Author based on US Census [2024b])

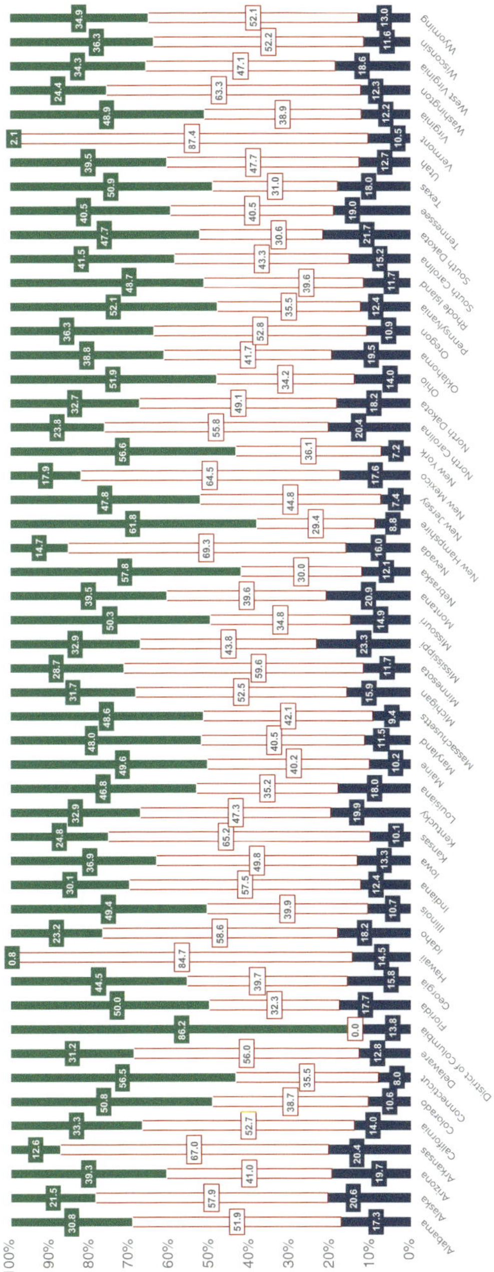

Chart 15.3 Percentage distribution of public elementary-secondary school system revenue by source and state: 2022 (*Source* Author based on US Census [2024b])

for this example. The program can aim higher and can be a gradual one that brings all states at an equal footing over a period of time. In Table 15.1, I show the basic calculation process starting from the per pupil spending in each state.

The logic of the table is simple and aims to equalise educational spending across states by bringing all states to the current average, $15,841 (Column A). To do this the table calculates the difference between the average and each state's per pupil expenditure (Column B). You can notice that some have a positive difference, and others a negative difference. This reflects the level of per pupil funding, above or below the average. Column C lists the number of pupils in each state. Column D multiplies the number of pupils by the difference (BxC). Column E provides the absolute value of Column D, and it removes all the negative signs.

Those who have a lower than the average spending level will get the difference from the federal government to boost their level of spending, and those who have a higher than the average spending level get the difference to alleviate state or local level funding. The point is to provide an across-the-board investment to upgrade educational spending levels.

The $168.8 billion yearly expenditure can be committed through a 6-year PCN that injects roughly $1 trillion into the US education system across states, recapitalising schools, endowing them with necessary technologies fit for the age of Artificial Intelligence, boosting skills and knowledge. The '6-year' does not refer to the maturity of the instrument (PCNs have no maturity), but to the deployment period. Naturally, this is a simplified calculation and a very basic example. My purpose is to offer a high-level justification for an Educational Equality PCN and not the entire structure and costing of the program.

Table 15.2 presents the US federal budget's discretionary outlays for 2023, where we can see the relative position of education. While the fifth highest allocation, education still lags far behind military spending, which is equal to 8.1 times the education spending. In other words, such an investment has ample justification. Both when looking at the past as well as the future, an Educational Equality PCN can fill a much observed, discussed, and debated issue in US public policy and society. A high space value opportunity can upgrade skills and opportunities across the board.

The point here is that an education-focused PCN can be used, alongside a NASA PCN, to transform the public expenditure architecture from debt to PCNs. This discussion aimed to show that through Value Easing, using Public Capitalisation Notes, a number of key public expenditure spheres can be allocated the necessary funding without increasing debt levels. Indeed,

Table 15.1 Total amount needed for educational equality PCN, 2022 figures in USD

	A Per pupil ($)	B Average—A	C Number of pupils 2022	D = BXC Additional Funding needed (+ or −)	E = |D| Additional Funding needed (+)
Alabama	11,819	4021.41	745,178	2,996,664,678.32	2,996,664,678.32
Alaska	20,191	−4350.09	129,944	−565,267,676.25	565,267,676.25
Arizona	10,315	5526.10	900,136	4,974,243,814.41	4,974,243,814.41
Arkansas	12,159	3681.74	465,413	1,713,531,621.90	1,713,531,621.90
California	17,049	−1208.38	5,348,692	−6,463,251,142.25	6,463,251,142.25
Colorado	13,422	2418.77	856,067	2,070,629,764.96	2,070,629,764.96
Connecticut	24,453	−8612.12	468,926	−4,038,445,501.48	4,038,445,501.48
Delaware	19,357	−3515.91	122,734	−431,521,144.34	431,521,144.34
District of Columbia	27,425	−11,583.87	48,635	−563,381,667.71	563,381,667.71
Florida	11,076	4764.95	2,805,486	13,368,002,399.21	13,368,002,399.21
Georgia	13,619	2222.18	1,699,426	3,776,435,114.59	3,776,435,114.59
Hawaii	17,420	−1579.59	173,178	−273,550,729.56	273,550,729.56
Idaho	9,670	6170.96	286,544	1,768,252,121.78	1,768,252,121.78
Illinois	18,927	−3085.73	1,861,261	−5,743,349,281.48	5,743,349,281.48
Indiana	12,322	3518.65	984,851	3,465,341,712.43	3,465,341,712.43
Iowa	13,259	2581.88	510,661	1,318,464,023.62	1,318,464,023.62
Kansas	14,408	1433.16	484,035	693,698,774.59	693,698,774.59
Kentucky	13,549	2291.48	654,098	1,498,849,823.38	1,498,849,823.38
Louisiana	14,928	912.51	607,687	554,520,522.85	554,520,522.85
Maine	17,885	−2043.68	170,093	−347,615,725.96	347,615,725.96
Maryland	17,753	−1912.02	881,064	−1,684,608,717.01	1,684,608,717.01
Massachusetts	21,906	−6064.91	859,323	−5,211,713,454.71	5,211,713,454.71
Michigan	15,719	122.27	1,244,451	152,159,197.55	152,159,197.55
Minnesota	15,441	400.20	797,739	319,252,154.83	319,252,154.83
Mississippi	10,984	4857.14	438,567	2,130,183,387.71	2,130,183,387.71
Missouri	12,631	3210.14	863,316	2,771,369,025.88	2,771,369,025.88
Montana	13,582	2259.06	148,977	336,548,062.35	336,548,062.35

(continued)

Table 15.1 (continued)

	A	B	C	D = BXC	E = |D|
	Per pupil ($)	Average—A	Number of pupils 2022	Additional Funding needed (+ or −)	Additional Funding needed (+)
Nebraska	14,285	1556.30	327,055	508,994,456.74	508,994,456.74
Nevada	11,677	4163.37	430,975	1,794,309,718.11	1,794,309,718.11
New Hampshire	21,605	−5764.16	160,371	−924,404,388.21	924,404,388.21
New Jersey	25,099	−9257.91	1,307,453	−12,104,282,098.57	12,104,282,098.57
New Mexico	13,260	2580.98	299,234	772,317,401.97	772,317,401.97
New York	29,873	−14,031.84	2,372,983	−33,297,316,124.91	33,297,316,124.91
North Carolina	12,298	3542.46	1,392,119	4,931,524,598.21	4,931,524,598.21
North Dakota	15,843	−1.78	116,834	−208,109.92	208,109.92
Ohio	15,583	257.45	1,554,712	400,267,560.08	400,267,560.08
Oklahoma	10,890	4951.28	638,821	3,162,979,285.91	3,162,979,285.91
Oregon	15,754	86.96	551,275	47,939,965.26	47,939,965.26
Pennsylvania	19,186	−3345.44	1,508,920	−5,048,000,259.48	5,048,000,259.48
Rhode Island	19,962	−4121.42	126,075	−519,607,892.49	519,607,892.49
South Carolina	13,387	2453.84	733,779	1,800,578,924.67	1,800,578,924.67
South Dakota	11,564	4276.63	141,021	603,094,326.40	603,094,326.40
Tennessee	11,317	4523.40	984,331	4,452,526,474.79	4,452,526,474.79
Texas	11,803	4037.96	5,053,217	20,404,712,179.44	20,404,712,179.44
Utah	9,552	6288.91	612,933	3,854,681,578.00	3,854,681,578.00
Vermont	24,608	−8767.14	83,675	−733,590,577.27	733,590,577.27
Virginia	15,059	782.05	1,249,688	977,320,685.12	977,320,685.12
Washington	17,119	−1278.48	1,075,813	−1,375,408,651.93	1,375,408,651.93
West Virginia	13,858	1982.97	252,207	500,119,143.00	500,119,143.00
Wisconsin	14,505	1335.90	818,471	1,093,399,483.34	1,093,399,483.34
Wyoming	18,529	−2688.62	92,756	−249,385,894.49	249,385,894.49
AVERAGE	**15,841**		910,612		3,309,565,118
Total				168,787,821,019.42	168,787,821,019.42

Source Author using US Census (2024b)

Table 15.2 USA federal budget discretionary actual outlays for 2023, billions

	2023 actual amount
Department of Defense—military	765
Department of Health and Human Services	145
Department of Veterans Affairs	131
Department of Transportation	98
Department of Education	94
Department of Homeland Security	77
Department of Housing and Urban Development	64
International assistance programs	49
Department of Energy	45
Other agencies	41
Department of Justice	36
Department of State	32
Department of Agriculture	31
National Aeronautics and Space Administration	25
Department of the Treasury	17
Department of the Interior	16
Department of Labor	14
Department of Commerce	11
Environmental Protection Agency	11
Social Security Administration	10
Corps of Engineers	9
Small Business Administration	2
Total	**1,722**

Source CBO (2023)

they can be achieved by reducing debt levels. Naturally, this depends on the insights and propositions made in earlier chapters.

The sectors that this can apply to are many. Given their high space value, transportation and health sectors are two obvious candidates for parallel PCNs. Ultimately, through their logic and application, PCNs can gradually transform the US Debt Ceiling into a Wealth Floor.

15.2 Conclusion

Adopting a space-adjusted financial value framework and mathematics can transform the monetary policy instruments we use for money creation. They can help us expand our toolkit and introduce the monetisation of space (PCN) alongside the monetisation of risk and time (debt) into our architecture. This can provide the tools through which we can fund our most ambitious projects in outer space. They can help us fund the programs we need to invent, build, deploy, and maintain the new habitats of the future. While this is the main proposition of this book, it is not the only benefit that can arise from such transformations.

One important additional benefit of this transformation is the potential capability to transform the US debt ceiling into a wealth floor. Through Value Easing and Public Capitalisation Notes we can also transform the architecture and instruments used to fund Federal Government expenditures. This can be done through parallel PCN programs, like an Educational Equality PCN that aims to invest in the US public education system and provide equal opportunities across states.

Value Easing and PCNs can gradually transform the debt ceiling into a wealth floor by shifting the nature of funding used. As mentioned, the use of PCNs and the resulting transformation can be a gradual process. Moreover, it does not need to be about new additional money. It can also start by changing the logic of creation for the amounts that are already being created.

References

Austin, A. 2015. The Debt Limit: History and Recent Increases. Congressional Research Service. https://sgp.fas.org/crs/misc/RL31967.pdf. Accessed 02 February 2020.

CBO. 2023. The Accuracy of CBO's Budget Projections for Fiscal Year 2023. Congressional Budget Office. https://www.cbo.gov/publication/59682#data. Accessed 12 June 2024.

Edelberg, W., and L. Sheiner. 2023. How Worried Should We be If the Debt Ceiling Isn't Lifted? Brookings. https://www.brookings.edu/articles/how-worried-should-we-be-if-the-debt-ceiling-isnt-lifted/. Accessed 12 July 2023.

IMF. 2024. Currency Composition of Official Foreign Exchange Reserves (COFER). The International Monetary Fund. https://data.imf.org/?sk=e6a5f467-c14b-4aa8-9f6d-5a09ec4e62a4. Accessed 12 June 2024.

Papazian, A. 2022. *The Space Value of Money: Rethinking Finance Beyond Risk and Time*. New York: Palgrave Macmillan. https://doi.org/10.1057/978-1-137-59489-1.

Papazian, A. 2023. *Hardwiring Sustainability into Financial Mathematics: Implications for Money Mechanics*. New York: Palgrave Macmillan. https://doi.org/10.1007/978-3-031-45689-3.

US Census. 2024a. Largest Year-to-Year Increase in Over 20 Years for Public School Spending Per Pupil. Press Release. https://www.census.gov/newsroom/press-releases/2024/public-school-spending-per-pupil.html. Accessed 12 June 2024.

US Census. 2024b. 2022 Public Elementary-Secondary Education Finance Data. US Census Bureau. https://www.census.gov/data/tables/2022/econ/school-finances/secondary-education-finance.html. Accessed 12 June 2024.

US Congress. 2021. Equality Act. United States Congress. https://www.congress.gov/bill/117th-congress/house-bill/5. Accessed 12 March 2024.

US Congress. 1974. H.R.40—Equal Educational Opportunities Act. United States Congress. https://www.congress.gov/bill/93rd-congress/house-bill/40. Accessed 12 May 2024.

US Treasury. 2023. Debt Limit. US Treasury, USA. https://home.treasury.gov/policy-issues/financial-markets-financial-institutions-and-fiscal-service/debt-limit. Accessed 12 January 2024.

US Treasury. 2024a. What is the national debt? US Treasury Fiscal Data. https://fiscaldata.treasury.gov/americas-finance-guide/national-debt/. Accessed 12 August 2024.

US Treasury. 2024b. Historical Debt Outstanding 1790–2023. US Treasury Fiscal Data. https://fiscaldata.treasury.gov/datasets/historical-debt-outstanding/historical-debt-outstanding. Accessed 12 June 2024.

US Treasury. 2024c. Total Public Debt Outstanding vs Debt Subject to Limit. US Department of the Treasury. https://www.treasurydirect.gov/news/home-page-articles-archive/release-11-08-04/#. Accessed 12 May 2024.

White House. 2023. The Potential Economic Impacts of Various Debt Ceiling Scenarios. The White House. https://www.whitehouse.gov/cea/written-materials/2023/05/03/debt-ceiling-scenarios/. Accessed 03 February 2024.

White House. 2024. Statutory Limits on Federal Debt: 1940–Current. White House. https://www.whitehouse.gov/omb/budget/historical-tables/. Accessed 12 June 2024.

16

UK New Outer Space Deal

I would like to be remembered as someone who was not afraid to do what she wanted to do, and as someone who took risks along the way in order to achieve her goals.
Sally Ride, STS-7 Challenger Astronaut, 1983

Never be limited by other people's imagination; never limit others because of your own limited imagination.
Mae Jameson, STS-47 Endeavour Astronaut, 1992

"I pledge you, I pledge myself, to a new deal for the American people." These were the words of Franklin D. Roosevelt (FDR) uttered in 1932, the promise that led to the New Deal he initiated as president. The New Deal was a series of programs and legislative initiatives that helped the US overcome the pains of the great depression and establish some of the structural components of the current banking and financial system in the US. "The Federal Emergency Relief Act of May 12, 1933, implemented President Roosevelt's first major initiative to combat the adverse economic and social effects of the Great Depression. The act established the Federal Emergency Relief Administration, a grant-making agency authorized to distribute federal aid to the states for relief. By the end of December 1935, FERA had distributed over $3.1 billion and employed more than 20 million people" (National Archives 2022). An amount roughly equivalent to $69 billion in 2023 US dollars (FEDM 2024).

© The Author(s), under exclusive license to Springer Nature
Switzerland AG 2024
A. V. Papazian, *Financing the Race to Space*,
https://doi.org/10.1007/978-3-031-73102-0_16

On the 16th of June 1933 FDR signed the Banking Act of 1933 into law, it established the Federal Deposit Insurance Corporation (FDIC), aimed at safeguarding depositors from the collapse of commercial banks (FDIC 2023). Part of the same package was the Glass-Steagall Act of 1933 which ensured that commercial banks cannot engage in investment banking activities in order to protect depositors from potential speculative losses in the stock market.[1] In 1934, as part of the same New Deal, the Securities Exchange Act established the U.S. Securities and Exchange Commission (SEC) which was entrusted with the regulation and enforcement of new securities laws aimed at addressing market manipulation and other types of securities fraud.

Similarly, on the 3rd of April 1948, US president Truman signed the Economic Recovery Act of 1948 into law, known as the Marshall Plan. "Over the next four years, Congress appropriated $13.3 billion for European recovery. This aid provided much needed capital and materials that enabled Europeans to rebuild the continent's economy" (National Archives 2022). In 2023 dollars, the Marshal Plan allocated roughly $168.5 billion to the European continent's recovery.[2]

The above brief historical parenthesis aims to remind the reader that fundamental improvements and changes in our monetary and financial architecture and spending allocations are very much possible. In fact, it is through such forward looking and bold institutional and policy actions that we have managed to make any progress at all.

This chapter argues that a New Outer Space Deal (NOSD) can be an umbrella legislative and institutional program that can mobilise the core insights and benefits of a space-adjusted financial value framework, mathematics, and monetary architecture. While a New Outer Space Deal can be designed and implemented in the US and elsewhere, it is my belief that the United Kingdom is ideally positioned and most in need of such a deal.

---

[1] The Glass-Steagall Act of 1933 was repealed in 1999, by the Gramm-Leach-Bliley Act (GLBA), the Financial Services Modernization Act of 1999. It removed the limitations on the securities activities of commercial banks and interactions between commercial banks and securities firms.

[2] Although not exactly the same kind of 'New Deals' as the above examples, in 2008 the Obama Administration tabled the Green New Deal, a term that was first coined by Thomas Friedman (Friedman, 2007); in 2009, the United Nations came up with its own Global Green New deal (United Nations, 2009a, 2009b); in 2019, the European Commission announced about the European Green Deal aimed at transforming the continent into a climate neutral continent by 2050 (EU, 2019); and again in 2019, the UK Labour party launched its own Green New Deal called the Green Jobs Revolution (UK Labour, 2019). More recently, in 2022, President Biden signed the Inflation Reduction Act into law, allocating $500 billion in new spending and subsidies aimed at boosting clean energy in the US (White House 2023).

16.1 Why the UK?

I have chosen to argue this case for the United Kingdom (UK) for a number of reasons. The first reason is the extensive policy and institutional work that has been done in the country through its 'National Space Strategy.' The second reason is that the UK has been at the front line of all outer space-related international legislation, and thus has all the required legislative background. The third reason is that the country has been plagued by a policy of austerity for 14 years now, and it is in dire need for new and properly funded economy wide investments. Finally, the fourth reason is the recently elected new Labour government that is determined to "rebuild Britain," and "take the breaks off Britain," which is exactly the kind of government such a plan will require. Moreover, the new government's manifesto and freshly released legislative agenda make no reference to outer space, which can be ideally rectified through a New Outer Space Deal.

16.1.1 National Space Strategy

The first most important reason is the National Space Strategy (NSS), which aims at transforming the UK into a science superpower and space power. As I have discussed throughout this book, I think this should have been called the *National Outer Space Strategy* (NOSS). Nevertheless, the interest and intention to boost the UK outer space sector reflect the right kind of background and context for this proposal.

The first policy paper on the National Space Strategy was published by the Department of Science, Innovation and Technology, the Ministry of Defence, and the Department of Business, Energy, and Industrial Strategy in 2021 (UK Government 2021). This policy paper was followed by an update in 2022 (UK Government 2022) and an Action report in 2023 (UK Government 2023). These strategy papers were complimented by the publication of the Defence Space Strategy in 2022 (UK MOD 2022), and the Space Industrial Plan in 2024.

This strategic commitment is exactly the kind of context that a New Outer Space Deal can plug into. In other words, the work that has been done to create the strategic and institutional components of a transformed UK outer space sector is the ideal backdrop that can be utilised to develop and launch a New Outer Space Deal to fund the country's greatest ambitions, with the kind of funding that allowed NASA to go to the moon in the 1960s.

The 2021 National Space Strategy policy paper established the vision.

We will build one of the most innovative and attractive space economies in the world, and the UK will grow as a space nation. We will protect and defend UK interests in space, shape the space environment and use space to help solve challenges at home and overseas. Through cutting-edge research, we will inspire the next generation and sustain the UK's competitive edge in space science and technology. (UK Government 2021, 6)

The 2022 updated national space strategy policy paper establishing the necessity for private investments and the willingness to remove barriers for growth.

Achieving the UK's goals in space will require coordinated action from government. However, public investment alone will not be sufficient. The UK will require a significant increase in private sector investment in space activities, and the full combined efforts of every participant in the UK space economy, from businesses to innovators, entrepreneurs, and space scientists. As the sector grows, develops, and matures over time, government will redefine our partnership with industry, changing from a primary funder to an influential customer. We will understand the barriers to growth and work with industry to reduce and remove them. (UK Government 2022)

The 2023 updated 'NSS in Action' policy paper discussed the ten-point plan and institutional progress.

Alongside delivering the 10 Point Plan, we have also made huge strides in improving the way we organise government to deliver on our space ambitions. DSIT and MOD have jointly brought together the responsible space departments into a single coordinated delivery framework overseen by a National Space Board. In July 2023, the Prime Minister established the National Space Council as a new Inter-Ministerial Group, to set cross-government ministerial direction for space policy and strategy. In addition, UKSA has outlined in its corporate plan how it is transforming into a world-leading delivery agency. As we enter the next phase of delivering the National Space Strategy, we will expand the range of activities we will focus on, delivering key enabling policies and interventions across every pillar of the National Space Strategy. Our joined-up approach to delivery will help us to identify and pursue opportunities for dual use capabilities and joint activity. (UK Government 2023)

The ten-point plan mentioned in the above quote consists of the following: (1) Capture the European market in commercial small satellite launch; (2) Fight climate change with space technology; (3) Unleash innovation across the space sector; (4) Expand our horizons with space science and exploration;

(5) Develop our world-class space clusters; (6) Lead the global effort to make space more sustainable; (7) Improve public services with space technology; (8) Deliver the UK Defence Space Portfolio; (9) Upskill and inspire our future space workforce, and (10) Use space to modernise and transform our transport system.

The implementation phases of the UK National Space Strategy are summarised in the 2022 paper and described in Fig. 16.1.

In 2023, in a report making 'the case for space,' the UK Government Department for Science, Innovation and Technology states:

> The UK space sector is worth over £17.5 billion in income to the UK economy and employs 48,800 people, offering some of the most productive and skilled jobs in the country. The sector has exhibited extremely robust growth since the turn of the millennium with an average long term growth rate of 6.4%, significantly outpacing growth of the wider UK economy and the global space economy (both of which grew at 1.6% on average per year over the period)… The UK plays an important part in the global space economy and is home to a healthy and growing space ecosystem. Boasting excellence across both business (with end-to-end capabilities throughout the space value chain) and academia (world-leader in fields of space science), the UK has cutting-edge expertise across multiple domains… *This is all despite the UK being a relatively low spender compared to the biggest space nations. The UK government spends around 0.05% of GDP on space, which is only around a fifth of that spent by the US (0.24%).* (UK DSIT 2023a, 4)[3]

Again in 2023, the UK Government Department for Science, Innovation and Technology also published its Science and Technology Framework where it identifies its broader vision.

> The motivation behind our Science and Technology Superpower agenda is simple: science and technology will be the major driver of prosperity, power and history-making events this century. The United Kingdom's future success as a rich, strong, influential country, whose citizens enjoy prosperity and security, and fulfilled, healthy and sustainable lives, will correspondingly depend on our ability to build on our existing strengths in science, technology, finance and innovation. (UK DSIT 2023b, 6)

In March 2024 the 'Space Industrial Plan' was released. In it we can read:

---

[3] Emphasis added.

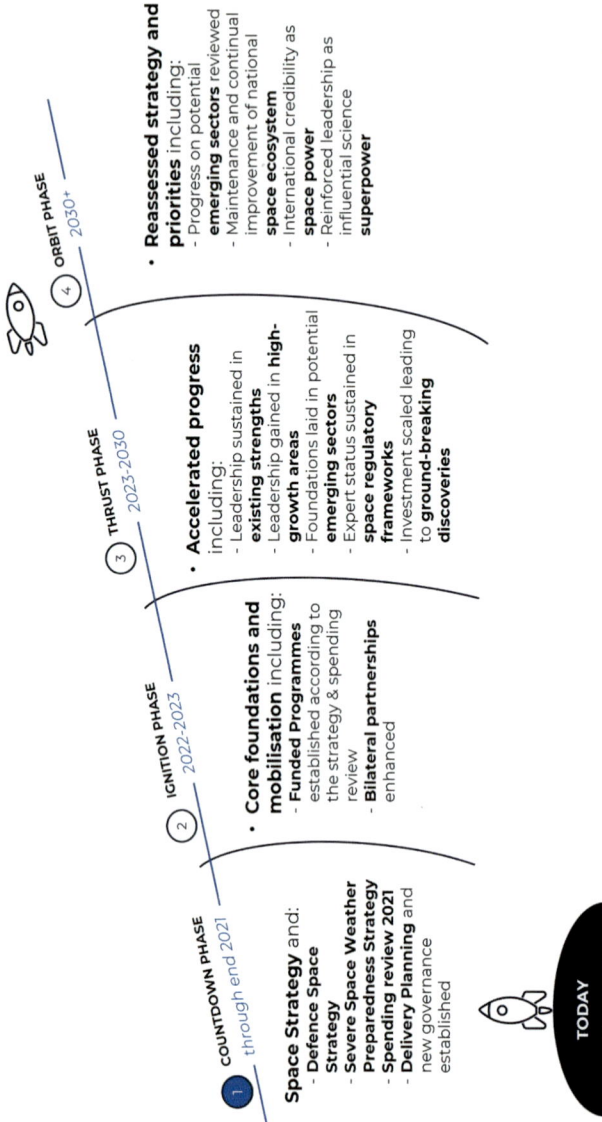

Fig. 16.1 UK National space strategy implementation phases (*Source* Adapted by author from UK Government [2022])

The timing of this publication is critical. The sooner we set out our priorities to the sector, the faster we can align our efforts and allow the UK to both seize emerging opportunities and address the evolving threats of an ever more congested and contested domain.... This publication coincides with a pivotal moment for how space activity is governed *as we enter the planning for the next Spending Review, and beyond. This will be the first full cycle in which the National Space Strategy, the Department for Science Innovation and Technology Space Directorate and Ministry of Defence Space Command have all been in place from the start, giving us an opportunity to design and deliver programmes and funding effectively across government.* (UK Government 2024a, 4)[4]

Last but not least, in a report published in the summer of 2024 by the National Audit Office, titled 'The National Space Strategy and the role of the UK Space Agency', we can read the following:

Space plays a critical role in modern everyday life in the UK. It is vital for scientific discovery and is a fast-developing UK commercial sector which has grown to around £17.5 billion in 2020-21. The government did well to draw its many different interests and activities in this very diverse sector into a single vision in its 2021 national Strategy, which set high ambitions and helped galvanise the sector's interest. DSIT recognised that the original Strategy was broad and that it did not know how much it would cost to deliver. However, it did not produce the implementation plan that it had originally planned to, and three years later DSIT and UKSA are still in the early stages of identifying and developing the plans and capabilities needed to deliver the Strategy's ambitions. (NAO 2024)[5]

The key findings of the NAO Report (NAO 2024) can be considered additional reasons why the UK is ready and would greatly benefit from a New Outer Space Deal. "The government has set out the UK's high-level approach to space, which has been welcomed by industry... The government restarted its cross-government ministerial council on space in July 2023, after a two-year hiatus... DSIT has not provided enough clarity or detail on its strategic ambitions to allow delivery bodies and stakeholders to plan to achieve them... DSIT does not yet fully understand the government's overall funding to and requirements for the space sector... DSIT has identified that the government's commercial approach to the sector will need to change, and it will need to develop capabilities across government in order to achieve this... UKSA has been proactive in working to align its activities

---

[4] Emphasis added.
[5] Emphasis added.

with the Strategy… UKSA and DSIT's process for allocating UKSA's £1.75 billion budget for 2022 to 2025 had some weaknesses but they are improving the approach for the next spending review period… The UK does not yet receive contracts from ESA proportionate to the value of the funding UKSA provides. UKSA is working with ESA to ensure this is the case by the end of December 2024… UKSA has not made as much progress as it planned on its programmes… UKSA wants to increase its use of alternative commercial interventions, such as procurement, to improve support for industry and attract more private investment, but it has not yet worked out the implications of this shift… DSIT and UKSA have identified that UKSA has more work underway than it can afford to continue, unchanged, beyond March 2025 without a budget uplift, and it may have to make difficult decisions on which of the Strategy's ambitions to prioritise… UKSA does not currently have a complete view of how it is progressing against its priorities, but it is working to develop one… At the Strategy level, DSIT is improving its view of progress but does not yet have a systematic framework for monitoring and evaluating progress across the whole Strategy…" (NAO 2024, 8–14).

Thus, the timing is also appropriate. The national space strategy, the defence space strategy, and the space industrial plan along with their relevant institutional and organisational components make the UK the ideal case for this New Outer Space Deal umbrella program—to boost and enhance long term investment in the sector, based on an entirely new and space-adjusted financial value framework, mathematics, and monetary tools.

Using a space-adjusted financial framework, the UK can allocate £15 billion a year to the UKSA, £75 billion over five years, with the objective to land the first British person on the moon by 2030, soon after the 2026 Artemis III mission "which is planned to land the first astronauts near the lunar South Pole" (NASA 2024). Indeed, this should be the 11th point of the ten-point plan revealed in the UK National Space Strategy and mentioned earlier. This is the kind of vision and investment the UK economy needs. In relative and absolute terms, lack of a better description, austere amounts cannot achieve grand visions in outer space.

16.1.2   Global Legislative Context

The second reason why the UK is an ideal case for a New Outer Space Deal has to do with its historic role in setting up the global legislative framework in relation to outer space. The existing global outer space legal framework is built around a number of key United Nations resolutions and treaties, and the UK has been at the centre of their development since the very beginning.

The story begins with the UN General Assembly Resolution 1721 (XVI) in 1961, which stated two key principles: "(a) International law, including the Charter of the United Nations, applies to outer space and celestial bodies; (b) Outer space and celestial bodies are free for exploration and use by all States in conformity with international law and are not subject to national appropriation" (UNOOSA 1961, 6). Since then, a number of global treaties have been approved. A good review of these regulations can be found in Butchard (2022) published as a research briefing of the UK House of Commons Library (see Table 16.1). Note that the UK is a signatory of the first four listed in the table.

Given the UK's early involvement, it was one of the three signatories of the Outer Space Treaty in 1967 alongside the United States and the Soviet Union (United Nations 1967), the country has all the relevant regulatory and legislative frameworks within and through which it can operationalise and deliver on the proposed 'New Outer Space Deal.' Indeed, this is one of the key strengths identified in the National Space Strategy as well. Furthermore, the UK is a signatory of the Artemis Accords, which has more signatories (50 by the 11th of December 2024) than the Moon Agreement has parties (17 as of December 2024).

> Grounded in the Outer Space Treaty of 1967 (OST), the Artemis Accords are a non-binding set of principles designed to guide civil space exploration and use in the 21st century. Co-led for the United States by the Department of State and the National Aeronautics and Space Administration (NASA), the Artemis Accords were launched on October 13, 2020 with Australia, Canada, Italy, Japan, Luxembourg, the United Arab Emirates, the United Kingdom and the United States. (USDOS 2024)

A New Outer Space Deal that allocates the right amount of funding that can take the UK to Orbit, just like NASA did in the 1960s, is a couple of legislations away.

16.1.3   Austerity and Underinvestment

The third reason why the UK is ideally positioned to implement this proposition is the history of fiscal austerity that has plagued the country and its public policy since 2010, since the post 2008 crisis period. After every Quantitative Easing program, in 2008/09 and 2020/21, the UK Government's policy of cutting budgets has been a defining theme and a subject of national debate.

Table 16.1 List of international outer space treaties

The Outer Space Treaty	The Outer Space Treaty provides the basic framework on international space law, covering legal foundations such as the peaceful use of space, the freedom of exploration of space, and the basic responsibility and liability of state for launching space objects	https://www.unoosa.org/oosa/en/our work/spacelaw/treaties/introouterspace treaty.html
The Rescue Agreement	The Rescue agreement provides that states shall take all possible steps to rescue and assist astronauts in distress and promptly return them to the launching state. It also provides that states shall, on request, provide assistance to launching states in recovering space objects that return to Earth outside the territory of the launching state	https://www.unoosa.org/oosa/en/our work/spacelaw/treaties/introrescueagre ement.html

The Liability Convention	This Convention provides for absolute liability on the part of a launching state to pay compensation for damage caused by its space objects on the surface of the Earth or to aircraft. It also makes launching states liable for damage in space, based on a fault of that state. The Convention also details provisions of the settlement of disputes	https://www.unoosa.org/oosa/en/our work/spacelaw/treaties/introliability-con vention.html
The Registration Convention	This Convention lays down the rules applicable for the registration of space objects, and the open and free access of these registers	https://www.unoosa.org/oosa/en/our work/spacelaw/treaties/introregistr ation-convention.html

(continued)

Table 16.1 (continued)

The Moon Agreement	This Agreement expands on the Outer Space Treaty specifically regarding the Moon and other celestial bodies. It provides that those bodies should be used exclusively for peaceful purposes, that their environments should not be disrupted, and that the UN should be informed about any station established on those bodies. The Agreement also provides that the Moon and its natural resources are the common heritage of mankind	https://www.unoosa.org/oosa/en/our work/spacelaw/treaties/intromoon-agr eement.html
Legally-binding treaty on preventing an arms race in space	At the Conference on Disarmament, Russia and China have been pushing for a legally-binding treaty on preventing an arms race in space since 2008, with an updated draft treaty proposed in 2014. As evidenced through the voting of General Assembly Resolution 76/23, the US, the UK and other aligned states do not support this initiative. It remains to be seen whether any progress will be made on this initiative, as the stalemate on developing the treaty seems to persist	https://www.nti.org/education-center/tre aties-and-regimes/proposed-prevention-arms-race-space-paros-treaty/

Source Adapted verbatim from Butchard (2022)

Chart 16.1 depicts the managed expenditure of the UK government and total public sector net debt excluding Bank of England. The managed expenditure total combines total public sector gross investment and total public sector current expenditure. For the year 2022/23, the £1,157.40 billion is made up of £106.34 gross investment and £ 1051.072 billion of current expenditure (OBR 2024). Chart. 16.2 depicts UK public sector net debt as a percentage of GDP based on Office for Budget Responsibility (OBR) figures. The percentage has reached 97.6% by 2022/23, from 63.9% in 1009/10.

We can clearly see the lost decade of austerity; years where public spending has stagnated, while public debts have risen. In Chart 16.3, we can observe UK's annual GDP and yearly growth rates. The austerity decade is sandwiched between the sharp declines caused by the 2007/2008 financial crisis and 2020 corona virus pandemic. The post-pandemic period has seen declining rates, and a sluggish recovery, combined with the supply shocks caused by the war in Ukraine and subsequent oil price increases and the cost-of-living crisis (see Chart 16.4). The above have been further exacerbated by the institutional challenges triggered by Brexit—UK's departure from the European Union which came into force on the 31st of January 2020 at midnight (CET).

In 2022, The Economist published an article with the title: "Britain is the sick man of Europe once again: The country will struggle to pep up its sluggish investment" (Economist 2022). In Chart 16.5, we can see that amongst the G7 countries, i.e., United States, Japan, Canada, Germany,

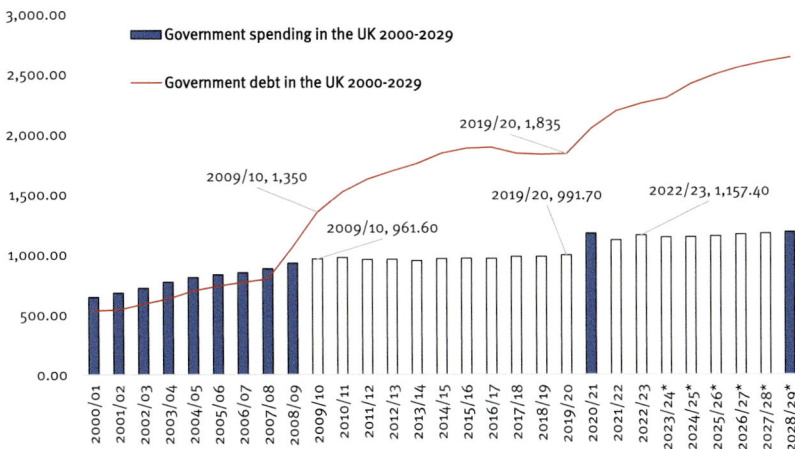

Chart 16.1   UK managed expenditure of the government and total public sector net debt excluding Bank of England from 2000/01 to 2028/29, in £billions (*Estimates) (*Source* Statista [2024a, 2024b])

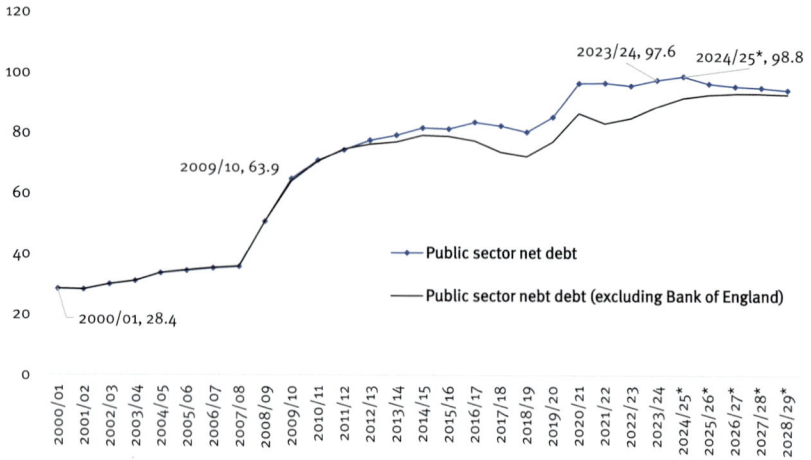

Chart 16.2 Public sector net debt expressed as a percentage of GDP in the UK (*Source* OBR [2024] and Statista [2024f])

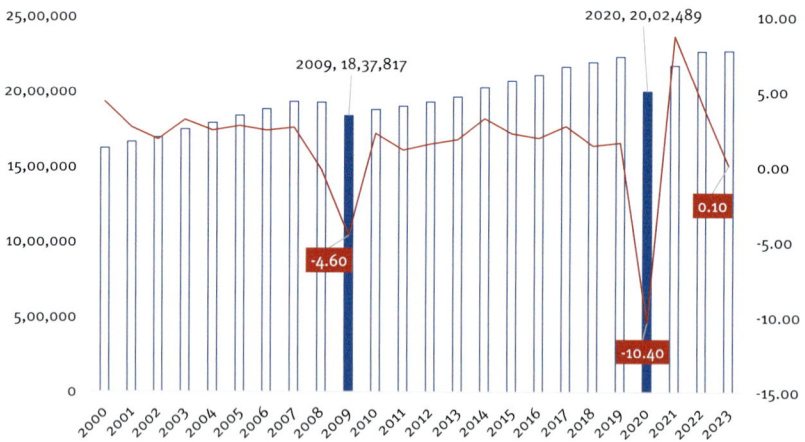

Chart 16.3 UK Gross Domestic Product (GDP) annual in £ millions and % growth (*Source* Statista (2024c, 2024d)

France, Italy, and the United Kingdom, in terms of Gross Fixed Capital Formation (GFCF), the UK was at the bottom of the group in 2011.

According to the World Bank (2024a), "Gross fixed capital formation (formerly gross domestic fixed investment) includes land improvements (fences, ditches, drains, and so on); plant, machinery, and equipment purchases; and the construction of roads, railways, and the like, including schools, offices, hospitals, private residential dwellings, and commercial and industrial buildings" (World Bank 2024b).

Chart 16.4 UK monthly Inflation rate—Consumer Price Index (CPI), 1989–2024, % (*Source* Statista [2024e])

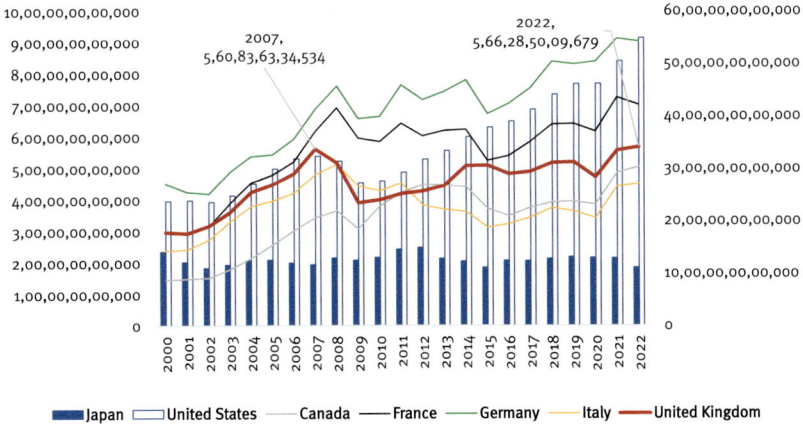

Chart 16.5 G7 gross fixed capital formation 2000–2022 (current US$) (*Source* World Bank [2024a])

The United States and Japan lead the group and due to the difference in the absolute figures are depicted on the right axis, all in current US Dollars. Germany and France are ahead of the UK, with the gap expanding after the 2007/8 financial crisis. The impact of austerity is evident in this chart. It took 15 years for UK's GFCF to reach its 2007 level. Looking at real wages in the same period, while they grew by 33% on average every decade from 1970 to 2007, they fell to below zero in the 2010s (Resolution Foundation 2022).

Within this context and background, I believe the UK is an ideal candidate for the proposed umbrella New Outer Space Deal. Indeed, the above discussion provides further background for the reader to assess the monetary commitment allocated to UK's National Space Strategy and its intentions to reach 'Orbit Phase' by 2030.

With the election of a new Labour government in July 2024, and the promise to address the decline in public investments, services, and infrastructure, this proposal is even more relevant. A change in framework is critical if the UK is going to manage a shift away from austerity.

16.1.4 Rebuilding Britain in Space

The fourth reason the UK is an ideal case for this proposition is its newly elected government that aims to rebuild Britain. On the 4th of July 2024, the British electorate voted for change and a new Labour Government came to power under the leadership of Sir Keir Starmer. On the 5th of July Reuters reported that "New PM Starmer pledges to rebuild Britain after years of chaos" (Reuters 2024a). Indeed, the five declared missions of the new government include: (1) kickstart economic growth; (2) make Britain a clean energy superpower; (3) take back our streets; (4) break down barriers to opportunity, and (5) build an National Health Service (NHS) fit for the future (UK Labour 2024a).

A few days later, on the 9th of July 2024, the new Chancellor of the Exchequer, Rachel Reeves, announced the creation of a National Wealth Fund that will invest in the new industries of the future (HM Treasury 2024a; GFI-OW 2024). The National Wealth Fund announcement reveals the strategic nature of this new venture, it aims to 'unlock private investment.' While there is a £7.3 billion commitment, the purpose is to 'crowd in' private investment.

The government's commitment to rebuild public services is evident, but the fiscal reality it faces remains challenging. Committed to sound money and fiscal discipline, the budgetary plans of the new government do not yet challenge the austerity budget and policy. Given our current financial value framework, mathematics, and monetary architecture, the government's options are limited.

The government does have a 'Green Prosperity Plan' that "will be funded in part by a time-limited windfall tax on the oil and gas giants making record profits, with the rest of the funding coming from responsible borrowing to invest within Labour's fiscal rules – catalytic investment that will leverage higher private investment and boost economic growth" (UK Labour 2024b).

While turning the page and putting an end to chaos, the new Labour government faces a daunting challenge. Unlocking growth and new investments that can address its economywide woes while relying on small public investments and expecting larger private investments may not be as effective as hoped. Naturally, much will be decided in the months and years to come, especially after the October 2024 new government budget.

However, what is absolutely clear is that outer space does not seem to be one of the five missions of the new Labour government, and the UK outer space industry is not at the forefront of the manifesto. As such, this proposition is an ideal opportunity to rectify and address this gap. Indeed, the King's Speech on the 17th of July 2024 outlined the new government's legislative agenda, where growth, energy, security, health, construction, and transportation are prominent, but there is no reference to outer space.

Table 16.2 lists the legislative agenda of the new Labour Government as announced in the King's Speech (UK Government 2024b). In the entire document, the word space is used twice, once referring to 'community spaces' and another time referring to 'public spaces.' There is absolutely no reference to outer space.

The first bill on the list is the Budget Responsibility Bill. In the introduction of the main agenda we can read: "Stability will be the cornerstone of my Government's economic policy and every decision will be consistent with its fiscal rules. It will legislate to ensure that all significant tax and spending changes are subject to an independent assessment by the Office for Budget Responsibility" (UK Government 2024b, 11). This is mainly designed to address the practice of unfunded tax cuts by previous governments. Nevertheless, it also indicates the necessity to rethink the fiscal and funding framework.

Prime Minister Keir Starmer has set a very clear agenda to "take the breaks off Britain." Given 14 years of austerity and a challenging fiscal landscape, taking the breaks off Britain may have to begin with the kind of innovations introduced in this book. Quoting Starmer:

> Now is the time to take the brakes off Britain. For too long people have been held back, their paths determined by where they came from - not their talents and hard work. I am determined to create wealth for people up and down the country. It is the only way our country can progress, and my government is focussed on supporting that aspiration. Today's new laws will take back control and lay the foundations of real change that this country is crying out for, creating wealth in every community and making people better off - supporting their ambitions, hopes and dreams. (UK Government 2024c)

Table 16.2 New labour government legislative agenda in King's speech 2024

Economic stability and growth

Budget Responsibility Bill
National Wealth Fund Bill
Pension Schemes Bill
Planning and Infrastructure Bill
Employment Rights Bill
English Devolution Bill
Passenger Railway Services (Public Ownership) Bill
Better Buses Bill
Railways Bill
Bank Resolution (Recapitalisation) Bill
Arbitration Bill
Product Safety and Metrology Bill
Digital Information and Smart Data Bill
High Speed Rail (Crewe to Manchester) Bill
Draft Audit Reform and Corporate Governance Bill
Great British energy and clean energy superpower
Great British Energy Bill
The Crown Estate Bill
Sustainable Aviation Fuel (Revenue Support Mechanism) Bill
Water (Special Measures) Bill
Secure borders, cracking down on anti social behaviour and take back our streets
Border Security, Asylum and Immigration Bill
Crime and Policing Bill
Terrorism (Protection of Premises) Bill
Victims, Courts and Public Protection Bill
Break down the barriers to opportunity
Children's Wellbeing Bill
Skills England Bill
Renters' Rights Bill
Football Governance Bill
Draft Leasehold and Commonhold Reform Bill
Draft Equality (Race and Disability) Bill
Draft Conversion Practices Bill
Health
Tobacco and Vapes Bill
Mental Health Bill
National security and serving the country
Hillsborough Law
Armed Forces Commissioner Bill
Northern Ireland Legacy Legislation

(continued)

Table 16.2 (continued)

Economic stability and growth
House of Lords (Hereditary Peers) Bill
Cyber Security and Resilience Bill
Commonwealth Parliamentary Association and International Committee of the Red Cross (Status) Bill
Lords Spiritual (Women) Act 2015 (Extension) Bill
Holocaust Memorial Bill
Holocaust Memorial Bill

Source UK Government (2024b)

Indeed, outer space can contribute to this new vision for the UK, and the strategic and institutional work has already been done through the National Space Strategy. A New Outer Space Deal is exactly what the UK needs, and what Starmer's plan can benefit from. *Rebuilding Britain in Space* will find domestic as well as international enthusiasm and resonance.

Overcoming austerity and increasing investment and productivity are directly and indirectly linked to the many limitations of our existing financial value framework and monetary architecture. As such, alongside the intention to go beyond austerity, we will need to take the brakes off Britain on the theoretical and framework level as well. Indeed, the new Chancellor of the Exchequer, Rachel Reeves, announced new budget cuts on the 29th of July aimed at addressing a £21.9 billion budgetary shortfall due to unfunded spending plans inherited from the previous government (Reuters 2024a, b). By October 2024, the New Labour budget (HM Treasury 2024) revealed the more complex reality facing the country and the government. The two key avenues through which the new Labour government aims to fund and achieve its vision of growth are higher taxes and more debts. The biggest tax increase in three decades is aimed to repair public finances and establish the foundations for future growth. Interestingly, while innovative technologies are a priority for the new government, the budget (HM Treasury 2024) makes no reference to the 'National Space Strategy' established in 2021. In fact, 'space' is hardly mentioned in the budget. Among the very few times that it is, the text refers to the allocation of "£975 million for the aerospace sector over 5 years to fund vital research and development for the latest aerospace technology."

16.2   UK New Outer Space Deal

A massive new investment program that raises the bar and adds more ambitious targets on the ten-point plan can be achieved through a space-adjusted financial value framework, mathematics, and monetary tools. If we were to transform our financial value framework and mathematics and introduce the concepts presented earlier, we will be able to allocate the billions needed to fund our most ambitious outer space projects. All this can be achieved while transcending austerity and the financial value framework and monetary architecture that justify it.

At the heart of a UK New Outer Space Deal can be a UK Space Agency—National Wealth Fund—Public Capitalisation Note (UKSA-NWF-PCN). I introduced the core concept and design of PCNs in previous chapters, giving the example of NASA. There is no reason why this instrument cannot be used by the UK to achieve the desired breakthroughs and fund a more ambitious National Outer Space Strategy (NOSS).

A UKSA-NWF-PCN can utilise the new institutional components in the public outer space sector and the new National Wealth Fund. Indeed, the entire vision and institutional capabilities established throughout the last few years can be empowered with appropriate funding to deliver on the scientific and space ambitions of the UK. Moreover, such a PCN powered program is in harmony with the private–public collaborative approach of the new government as well as the 'connect to prosper' agenda of the 695th Lord Mayor of London, Alderman Professor Michael Mainelli (Mainelli 2023).

As discussed in previous chapters, this is directly linked to our financial value framework and monetary architecture. In our current reality, a change of framework is required before we can operationalise a new policy that goes beyond austerity. Debt-based fiscal discipline can be replaced with a PCN-based fiscal discipline. This is important to note as the suggestion is not to become fiscally reckless. The idea is to change the framework of how that discipline is managed.

Figure 16.2 offers the structure and design that can be used to fund UK's New Outer Space Deal. The non-debt high space value features of the instrument will ensure that the issuance of a PCN does not affect the debt levels of the country. Moreover, given that the new money is injected directly in productive activity, the likelihood of asset price inflation is actually much lower than in previous debt-based injections.

Most importantly, a PCN funded New Outer Space Deal can allow the government to aim higher, like land the first British person on the moon. It will also allow the delivery of its most ambitious visions without having to

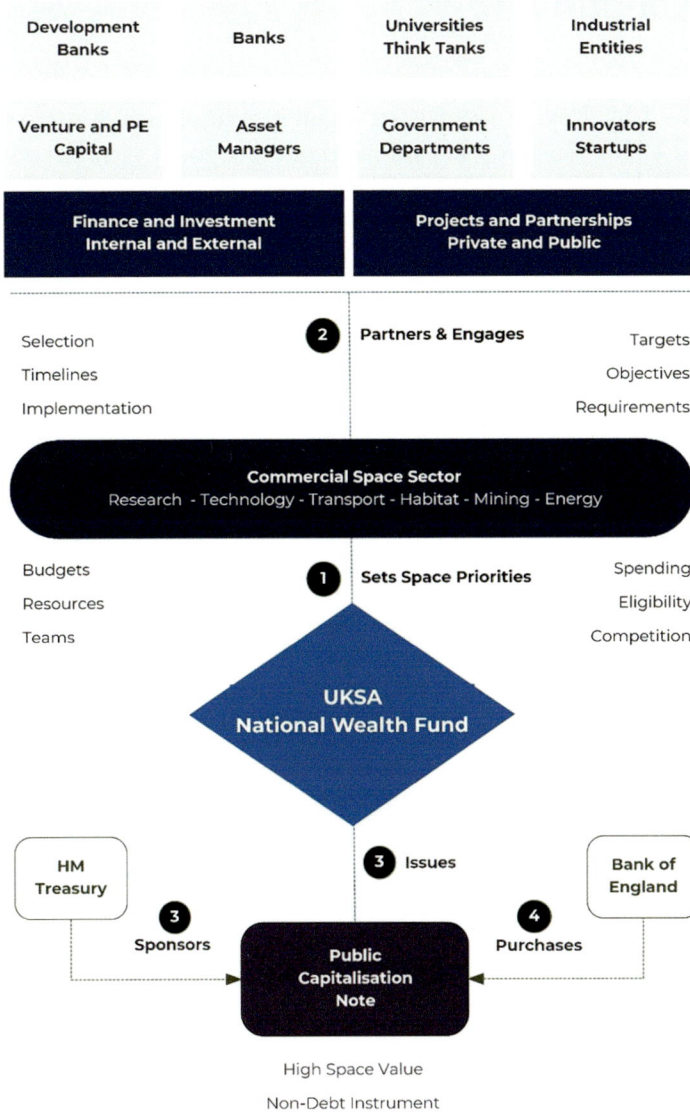

Fig. 16.2 UKSA NWF PCN (*Source* Adapted and updated from Papazian [2022, 2023])

cope with the constraints of a meagre budget (£10 billion over a decade) after 14 years of austerity. The positive impact of a UKSA NWF PCN will be felt across the technology startup community, existing space industry, manufacturing sector, as well as the education sector. It will also lead to higher tax revenues down the line, and a more dynamic environment liberated from the debilitating and visionless influence of austerity.

16.3 Conclusion

Austerity for future prosperity is a commonly revisited and experienced theme in the UK. J.M. Keynes spent years arguing against such absurdities in the 1920s and 1930s. Austerity as national economic policy is in fact a symptom of a distorted sense of conservatism and deep institutional risk aversion. It seems that UK's cultural inclination towards gentle evolution have landed the country in economic stagnation. A policy cul-de-sac that must be addressed on the theoretical and framework level first.

Since its official announcement in 2010, austerity, under the guise of 'fiscal discipline,' has led to consistent increases in debt and tax levels while eroding and defunding many public services through years of underinvestment. Indeed, while investment decline has been achieved, public finances have not improved. The new Labour Chancellor of the Exchequer, Rachel Reeves, described the situation as "worst state since second world war" (Financial Times 2024). Austerity has led to nothing more than the increased need for large public investments to upgrade and increase domestic investment and productivity.

As such, the United Kingdom is ideally positioned to implement a New Outer Space Deal. With the development of the National Space Strategy in 2021 and the progress achieved since then, the UK can implement a New Outer Space Deal and trigger a massive investment program that can raise the bar and funding of the strategy and put the first British person on the moon by 2030. When there is enough funding and a national mission, it is possible to conceive and implement the technological and collaborative partnerships that can make this happen. Naturally, the benefits will be felt across industries and the entire nation.

On the 22nd of July 2024 the Department for Science, Innovation and Technology announced a £33 million boost for the national space programme (UK Government 2024d). £24 million of the total amount is allocated to 8 different projects, which include a sub-orbital rocket test, micro reactors to support space exploration, and a heat-detecting telescope to gather data that can be used to help tackle the climate crisis. While the strategic commitment is present, and the projects are all worthwhile, the amounts and the ambition is still modest. Meanwhile, on July the 2nd, Roscosmos announced and shared its roadmap to building a new space station at a cost of $7 billion (Space 2024), with the first stage to launch in 2027.

Considering relative GDP figures (IMF 2024), there is no reason why the 6th largest economy in the world cannot do just as much if not more than the 11th largest economy. Recognising and accounting for differences in past and recent investments in the sector, this reinforces the need and necessity for greater ambition and commensurately greater levels of funding for the UK outer space sector.

A New Outer Space Deal built on the transformations proposed in this book can address the theoretical and policy quagmire that austerity, high taxes, and high debt levels have created. Although, as suggested in the previous chapter, this can also be done to address and rectify the lack of investment in the public health sector as well. As such, a 'New Health Deal' that injects relevant resources in the National Health Service is also possible. In fact, given a parallel crisis in public transportation, a 'New Transportation Deal' can also be put in place.

Given the circumstances, and the discussion in this book, to "take the breaks off Britain," to unshackle the country from the debilitating grip of austerity, we will have to seek the theoretical and framework transformations first. Breaking away from years of unproductive stagnation will require a rethink of the theoretical framework and monetary architecture that underpin the domestic policy matrix.

References

Butchard, P. 2022. International Regulation of Space. House of Commons Library Research Briefings. https://researchbriefings.files.parliament.uk/documents/CBP-9432/CBP-9432.pdf. Accessed 12 May 2024.

Economist. 2022. Britain Is the Sick Man of Europe Once Again: The Country Will Struggle to Pep Up Its Sluggish Investment. The World Ahead in 2023. The Economist. November 2022. https://www.economist.com/the-world-ahead/2022/11/18/britain-is-the-sick-man-of-europe-once-again. Accessed 12 December 2023.

EU. 2019. The European Green Deal Sets Out How to Make Europe the First Climate-Neutral Continent by 2050. European Commission. https://ec.europa.eu/commission/presscorner/detail/en/ip_19_6691. Accessed 12 May 2024.

FDIC. 2023. U.S. Banking and Deposit Insurance History: 1930–1939. Federal Deposit Insurance Corporation. https://www.fdic.gov/about/history/deposit-insurance/1930-1939.html. Accessed 12 May 2024.

FEDM. 2024. Inflation Calculator. Federal Reserve Bank of Minneapolis. https://www.minneapolisfed.org/about-us/monetary-policy/inflation-calculator. Accessed 12 June 2024.

Financial Times. 2024. Rachel Reeves Warns UK Public Finances in Worst State since Second World War. Financial Times. https://www.ft.com/content/ab7a74a0-3353-4254-b112-b8fd19bf5bd2. Accessed 08 July 2024.

Friedman, T. 2007. The Power of Green. The New York Times. https://www.nytimes.com/2007/04/15/opinion/15iht-web-0415edgreen-full.5291830.html. Accessed 12 May 2024.

GFI-OW. 2024. National Wealth Fund Taskforce. Green Finance Institute and Oliver Wyman. https://www.greenfinanceinstitute.com/wp-ontent/uploads/2024/07/20240709_1400_NWF-Taskforce-Report-v.FINAL_.pdf. Accessed 09 July 2024.

HM Treasury. 2024. National Wealth Fund: Mobilising Private Investment. UK Government. HM Treasury. https://assets.publishing.service.gov.uk/media/6710cf42080bdf716392f558/NWF_IIS_Publication.pdf. Accessed 02 November 2024.

HM Treasury. 2024b. Autumn Budget 2024: Fixing The Foundations to Deliver Change. HM Government. https://assets.publishing.service.gov.uk/media/6722120210b0d582ee8c48c0/Autumn_Budget_2024__print_.pdf. Accessed 10 November 2024.

IMF. 2024. GDP, Current Prices, World, Billions of U.S. dollars. International Monetary Fund. https://www.imf.org/external/datamapper/NGDPD@WEO/OEMDC/ADVEC/WEOWORLD. Accessed 24 July 2024.

Mainelli, M. 2023. Connect to Prosper. City of London Corporation. https://www.cityoflondon.gov.uk/about-us/about-the-city-of-london-corporation/lord-mayor/connect-to-prosper. Accessed 12 December 2023.

Mills, C. 2021. The Militarisation of Space. House of Commons Library Research Briefings. https://researchbriefings.files.parliament.uk/documents/CBP-9261/CBP-9261.pdf. Accessed 12 May 2024.

NASA. 2024. NASA Shares Progress Toward Early Artemis Moon Missions with Crew. NASA. https://www.nasa.gov/news-release/nasa-shares-toward-early-artemis-moon-missions-with-crew/. Accessed 12 May 2024.

National Archives. 2022. Marshall Plan (1948). National Archive Milestone Documents. https://www.archives.gov/milestone-documents/marshall-plan. Accessed 12 May 2024.

NAO. 2024. The National Space Strategy and the role of the UK Space Agency. National Audit Office. https://www.nao.org.uk/wp-content/uploads/2024/07/national-space-strategy-and-the-role-of-the-uk-space-agency.pdf. Accessed 12 September 2024.

OBR. 2024. The Evolution of Public Sector Net Debt (Excluding the Bank of England) Since 2000/. Office for Budgetary Responsibility. UK Government. https://obr.uk/box/the-evolution-of-public-sector-net-debt-excluding-the-bank-of-england-since-2000/. Accessed 12 June 2024.

Papazian, A. 2023. *Hardwiring Sustainability into Financial Mathematics: Implications for Money Mechanics*. New York: Palgrave Macmillan. https://doi.org/10.1007/978-3-031-45689-3.

Papazian, A. 2022. *The Space Value of Money: Rethinking Finance Beyond Risk and Time*. New York: Palgrave Macmillan. https://doi.org/10.1057/978-1-137-594 89-1.

Resolution Foundation. 2022. Stagnation Nation: Navigating a Route to a fairer and more prosperous Britain, Resolution Foundation & Centre for Economic Performance, LSE. https://www.nuffieldfoundation.org/wp-content/uploads/2022/07/Stagnation-nation-interim_report.pdf. Accessed 12 June 2024.

Reuters. 2024a. New PM Starmer Pledges to Rebuild Britain After Years of Chaos. https://www.reuters.com/world/uk/new-pm-starmer-pledges-action-not-words-fix-britain-2024-07-05/. Accessed 06 July 2024.

Reuters. 2024b. New UK Government Cuts Billions of Pounds of Spending to Fix 'Unsustainable' Finances. https://www.reuters.com/world/uk/new-uk-finance-minister-accuse-last-government-multi-billion-pound-cover-up-2024-07-28/. Accessed 30 July 2024.

Space. 2024. Russia Unveils Timeline for Building Its New Space Station, Starting in 2027. Space.com. https://www.space.com/russia-space-station-timeline-2027. Accessed 24 July 2024.

Statista. 2024a. Total managed expenditure of the government of the United Kingdom from 2000/01 to 2028/29. https://www.statista.com/statistics/298465/government-spending-uk/. Accessed 12 June 2024.

Statista. 2024b. Total public sector net debt excluding Bank of England in the United Kingdom from 2000/01 to 2028/29. https://www.statista.com/statistics/282647/government-debt-uk/. Accessed 12 June 2024.

Statista. 2024c. Gross domestic product of the United Kingdom from 1948 to 2023. https://www.statista.com/statistics/281744/gdp-of-the-united-kingdom/. Accessed 12 June 2024.

Statista. 2024d. Annual growth of gross domestic product in the United Kingdom from 1949 to 2023. https://www.statista.com/statistics/281734/gdp-growth-in-the-united-kingdom-uk/#. Accessed 12 June 2024.

Statista. 2024e. Inflation rate for the Consumer Price Index (CPI) in the United Kingdom from January 1989 to April 2024. https://www.statista.com/statistics/306648/inflation-rate-consumer-price-index-cpi-united-kingdom-uk/. Accessed 12 June 2024.

Statista. 2024f. Public sector net debt expressed as a percentage of GDP in the United Kingdom from 1900/01 to 2028/29. https://www.statista.com/statistics/282841/debt-as-gdp-uk/. Accessed 12 June 2024.

UK DSIT. 2023a. The Case for Space: Investing to realise its potential for UK benefit. Uk Government Department of Science, Innovation and Technology. https://assets.publishing.service.gov.uk/media/64afdb40c033c1001080623b/the_case_for_space.pdf. Accessed 12 May 2024.

UK DSIT. 2023b. Science and Technology Framework. UK Government Department of Science, Innovation and Technology. https://assets.publishing.service.gov.uk/media/6405955ed3bf7f25f5948f99/uk-science-technology-framework.pdf. Accessed 12 May 2024.

UK Government. 2024a. Boost for new National Wealth Fund to unlock private investment. https://www.gov.uk/government/news/boost-for-new-national-wealth-fund-to-unlock-private-investment_. Accessed 09 July 2024.

UK Government. 2024b. King's Speech 2024: background briefing notes. UK Government. https://assets.publishing.service.gov.uk/media/669791c549b9c05 97fdafe63/King_s_Speech_2024_background_briefing_notes.pdf. Accessed 17 July 2024.

UK Government. 2024c. King's Speech to unlock growth and "take the brakes off Britain". UK Government. https://www.gov.uk/government/news/kings-speech-to-unlock-growth-and-take-the-brakes-off-britain. Accessed 17 July 2024.

UK Government. 2024d. £33 million boost for national space programme. https://www.gov.uk/government/news/33-million-boost-for-national-space-programme. Accessed 23 July 2024.

UK Government. 2023. National Space Strategy in Action. https://www.gov.uk/gov ernment/publications/national-space-strategy-in-action/national-space-strategy-in-action#delivery-of-the-national-space-strategy. Accessed 12 April 2024.

UK Government. 2022. National space strategy. https://www.gov.uk/government/ publications/national-space-strategy/national-space-strategy#part-3-how-the-uk-will-achieve-our-goals. Accessed 12 April 2023.

UK Government. 2021. National space strategy. https://assets.publishing.service. gov.uk/media/6196205ce90e07043d677cca/national-space-strategy.pdf. Accessed 12 April 2023.

UK labour. 2024a. Mission-driven government. UK Labour Party. https://labour. org.uk/change/mission-driven-government/. Accessed 06 July 2024.

UK Labour. 2024b. Labour's fiscal plan. UK Labour Party. https://labour.org.uk/ change/labours-fiscal-plan/. Accessed 06 July 2024.

UK Labour. 2019. A Green Jobs Revolution. Labour for a New Green Deal. UK Labour Party. https://static1.squarespace.com/static/5c742a3c77b9036ccae1e ddf/t/60a77759442557630dc8dcbd/1621587821008/LGND_21_CONF_REP ORT_-compressed.pdf. Accessed 12 March 2024.

UK MOD. 2022. Defence Space Strategy: Operationalising the Space Domain. Ministry of Defence, UK Government. https://assets.publishing.service.gov. uk/media/61f8fae7d3bf7f78e0ff669b/20220120-UK_Defence_Space_Strategy_ Feb_22.pdf. Accessed 12 April 2024.

United Nations. 2009a. The Global Green New Deal. United Nations Department of Economic and Social Affairs. https://sdgs.un.org/publications/global-green-new-deal-17506. Accessed 12 May 2024

United Nations. 2009b. UN-DESA Policy Brief No. 12: A Global Green New Deal for Sustainable Development. United Nations Department of Economic and Social Affairs. https://sdgs.un.org/sites/default/files/publications/policybri ef12.pdf. Accessed 12 June 2024.

United Nations. 1967. Treaty on Principles Governing the Activities of States in the Exploration and Use of Outer Space, including the Moon and Other Celestial Bodies. https://treaties.unoda.org/t/outer_space. Accessed 12 May 2024.

UNOOSA. 1961. UN Resolution 1721 (XVI): International Co-operation in the Peaceful Uses of Outer Space. https://www.unoosa.org/pdf/gares/ARES_16_1 721E.pdf. Accessed 12 April 2024.

White House. 2023. Inflation Reduction Act Guidebook. The White House. https://www.whitehouse.gov/cleanenergy/inflation-reduction-act-guidebook/. Accessed 12 March 2024.

USDOS. 2024. Artemis Accords. United States Department of State. https://www.state.gov/artemis-accords/#. Accessed 24 June 2024.

World Bank. 2024a. Gross fixed capital formation (current US$). The World Bank Database. https://databank.worldbank.org/reports.aspx?source=2&series=NE.GDI.FTOT.CD&country=#. Accessed 12 June 2024.

World Bank. 2024b. Gross Fixed Capital Formation. The World Bank Data Glossary. https://databank.worldbank.org/metadataglossary/world-development-indicators/series/NE.GDI.FTOT.CD. Accessed 12 June 2024.

17

Conclusion

We are infinite beings with infinite possibilities. We have the power within ourselves to do anything we set our minds to.
Bernard A. Harris Jr., STS-55 Columbia Astronaut, 1993

Looking up, I see the immensity of the cosmos; bowing my head, I look at the multitude of the world. The gaze flies, the heart expands, the joy of the senses can reach its peak, and indeed, this is true happiness.[*]
Samantha Cristoforetti, Soyuz TMA-15 M Futura and SpaceX Crew-4 Astronaut, 2014

[*]*A part translation of the 'Preface to the Poems Collected from the Orchid Pavilion'), or Lanting Xu – "Orchid Pavilion Preface", a piece of Chinese calligraphy considered to be written by Wang Xizhi (303–361).*

We must change ourselves and our value framework to be able to truly change course. Given the evidence and analysis provided in this book, staying on our current trajectory is equivalent to living in an interpretation of human civilisation that makes us indistinguishable from a planet-consuming and space-littering parasite. However proud we may be of our technological achievements and our progress to date, and however, inspired we may feel by our national, cultural, and/or religious heritage, this is a fact.

While it is hypothetically possible for us to extend our reach into the cosmos without the changes discussed in this book, in our current scenario and given the levels of investment in outer space, we will struggle to secure our continuity on another planet and/or habitat. Even if we were to accept

A. V. Papazian, *Financing the Race to Space*,
https://doi.org/10.1007/978-3-031-73102-0_17

and yield to the planet-consuming space-littering interpretation of human civilisation, we are not expanding fast enough in outer space to secure the evolutionary continuity of the parasite.

Our inability to invest and create a sustainable and fair reality on earth and our inability to invest and expand our reach in outer space are dragging us into a trajectory of ecological destruction and eventual oblivion. In other words, we are not expanding in outer space at the speed with which we can secure an alternative future for humanity, while we are destroying our only home at alarming speed. Indeed, we are mistreating space in the broadest sense of the word—our physical context of matter, irrespective of constitution, composition, density, dynamics, and temperature, stretching from subatomic to interstellar space and every layer in between and beyond.

To be able to invest and build sustainably and fairly on Earth, to be able to invest and create that which we need to build to become a multiplanetary species, or a multi-habitat species, we must change and reform our financial value framework and reinvent the very principles that define the value of money and its creation. Anything short of such a fundamental change may lead to momentary breakthroughs, may create a growing Earthbound outer space industry, and may trigger the establishment of extensive and exhausting sustainability standards, but will fall short of providing us the monetary and financial foundations necessary for our sustainable and continuous expansion and settlement in outer space.

Our entire analytical framework in finance is built around risk and time, without space, based on two principles of value, Risk and Return and Time Value of Money, which discriminate against our evolutionary investments due to their biases against highly risky and very distant cash flows. Moreover, our framework is designed to serve only one stakeholder, the mortal risk-averse return-maximising investor and her/his/their preferences.

The omission of space as an analytical dimension in finance has led to the abstraction of our responsibility for space impact, and thus the abstraction of space impact from our models. This is particularly relevant to evolutionary investments, such as outer space exploration and settlement, given that the risk and time features of such investments are at odds with our principles and equations of value in finance.

Looking closely at the outer space economy, we observed that the species spends more on trying to sell goods and services to itself than on building its collective future in outer space. Moreover, the species spends more on defending itself from itself than on building its collective future in outer space. The private outer space economy, given its pursuit of Earthly money supply is Earthbound. While not constrained with such requirements, the public outer space economy is bound to Earth through its own funding sources, governmental budgets, and debts.

All these limitations can be traced back to our financial value framework, financial mathematics, and monetary architecture, and this book discussed and explored why this is so and how we can transcend these constraints. Indeed, humanity must be able to fund massive investments in outer space education, R&D, and manufacturing, if it is going to be able to sustain lunar habitats and other exploration missions beyond LEO, MEO, and GEO. Outer space exploration, development, and settlement is an evolutionary challenge for humanity, and it is directly affected by a spaceless financial value framework and monetary architecture that bind us to risk and calendar time, thwarting the timeless and hazardous investments we must initiate to break through.

Space is critical for balancing our time and risk focused financial value framework and equations, and necessary to allow the positive valuations of opportunities that are highly risky, have distant cash flows, and imply significant space impact. As a direct consequence of our financial value framework and mathematics, space is also absent from our monetary architecture, where money is created through debt instruments designed in and valued through a risktime framework without space and outer space.

Three key architectural challenges created by our debt-based monetary architecture impose serious systemic limitations on our ability to invest and explore space and outer space. Using calendar time as a foundational pillar of money creating instruments chains everyone to calendar time and acts as a muzzle on our ability to invest in space timelessly. Using debt instruments linked to calendar time acts as a leash on the species, due to the backward loop included in the structure of the instruments, limiting the distance we can travel in space before having to return to some bank, ultimately chaining us to the surface of the planet. Using debt instruments for money creation, whatever the level of capital accumulation, creates an artificial chase for cash and deposits, a monetary hunger which coupled with the threat of default, and loss of assets and ratings, acts as a whip in space.

Given our current financial value framework and monetary architecture, our entire monetary potential and thus productive capacity is chained to risk and calendar time, to the rotation and revolution of the Earth, to the surface of the planet, triggered to consume it at all costs, to serve a debt-based monetary architecture that has primacy over our ecosystem, over space and outer space.

I proposed the introduction of the analytical dimension of space into finance, along with the space value of money principle and ensuing equations, to rectify the structural imbalance of our risk and time focused framework. A framework that is currently crippling our ability to invest in and address our many evolutionary challenges, which include outer space exploration, planetary sustainability, and numerous socioeconomic crises (Papazian, 2022, 2023).

To enable and empower our outer space ambitions and address our evolutionary challenges, we must introduce the analytical dimension of space and the associated principle of value, the space value of money, into finance theory and practice. The space value of money states that a dollar invested in space must, at the very least, have a dollar's worth of positive impact on space. This allows the relevant and necessary transformations in our financial mathematics, where space-adjusted equations can facilitate both the sustainability of our activities on Earth as well as the funding and expansion of our footprint in outer space.

The space value of money principle and framework entrench respect for space and establish the foundations upon which we can integrate our space impact into our equations of value and return. This leads to a financial mathematics that values space, in parallel to and beyond risk and time. Our investments with very high risks and very distant cash flows, like outer space exploration projects, can now have a positive value independently of their risk and time profile.

We must be able to integrate the space and outer space impact of cash flows into the valuation of cash flows to be able to balance their risk and time value. The discounting of future expected cashflows must be accompanied by the compounding of their space impact into the future, if our monetary and financial system is to support our ambitions. Our footprint must create its value in outer space, without linking it to the concerns of individual mortal beings on Earth, i.e., risk and time, and without conditioning it by the planet's rotation on itself and revolution around the sun. This is necessary if the outer space industry is to grow beyond Earthbound services, and beyond Earthly money supply.

Following the introduction of the principle and equations of the space value framework, the next necessary step is to transform money creation based on the principle and equations that provide the blueprint of a new form of monetisation, Value Easing. Using a non-debt, no-maturity, high space value, equity-like instruments like Public Capitalisation Notes, we can transcend the limitations imposed by debt-based money. Any species that drives and guides its own creativity and productivity through money and monetary incentives must integrate space, our physical context, into its equations of value and return, if it is to enable long horizon high risk evolutionary investments into its economic, monetary, and financial fabric. We must monetise space to explore it. Our current money creation instruments, i.e., debts, monetise risk and calendar time.

Once such an alternative channel of money creation is introduced, we will be able to create and allocate the necessary and required levels of funding for outer space exploration and settlement. This transformation will also allow us to transcend the US debt limit and transform it into a wealth floor. Another parallel benefit is the design and implementation of a New Outer Space Deal which can establish relevant legislative foundations and mobilise resources to empower the outer space sector.

I believe the United Kingdom, given years of austerity as well as the recently published National Space Strategy, is the ideal locale for the implementation of such a proposition. With a new Labour government focused on growth and on turning the page, a UK New Outer Space Deal can be exactly the platform that can inject the much-needed investments in the UK economy. Unshackling the country from the debilitating grip of austerity requires a rethink of the theoretical and mathematical framework that underpins the assumptions and understanding of fiscal discipline.

To secure the sustainability as well as future expansion of human productivity, we must be able to respect, value and explore space and its many layers. To achieve such a reconfiguration, a civilisational leap from where we are today, we must reinvent our financial value framework, mathematics, and monetary architecture. The space value framework provides the theoretical and mathematical foundations through which we can engineer the new products and instruments that transform a debt-based monetary system and allow us to invest in our evolutionary challenges. Outer space development, exploration, and settlement is one such challenge and opportunity.

The space value of money principle, the financial mathematics that ensues, and the monetary transformations it can lead to entrench the respectful treatment of space into our models, allow the integration of our footprint, and empower us to create value where we stand, and wherever we may go, irrespective of calendar time and the risks involved. And if we were to one day transcend money, and mobilise human creativity and productivity without monetary incentives, we would still need to start with respect. Respecting and valuing space and its many layers are the milestones necessary to ensure our sustainability on Earth and expansion in outer space.

The cosmic landscape we find ourselves entangled with is fused with our own interpretations of our context. As such, the ultimate key to any and all progress must first be found in our own imagination.

References

Papazian, A. 2022. *The Space Value of Money: Rethinking Finance Beyond Risk and Time*. New York: Palgrave Macmillan. https://doi.org/10.1057/978-1-137-594 89-1.

Papazian, A. 2023. *Hardwiring Sustainability into Financial Mathematics: Implications for Money Mechanics*. New York: Palgrave Macmillan. https://doi.org/10. 1007/978-3-031-45689-3.

Index